AF262839

The Good Building Book

OPPOSITE Foxes & cherries, Electric
Avenue, Brixton, by Lucy Casson.

The Good Building Book

Principles of Efficient, Functional and Sustainable Design

Jon Broome and Nick Grant

GREEN BOOKS

LONDON · OXFORD · NEW YORK · NEW DELHI · SYDNEY

GREEN BOOKS
Bloomsbury Publishing Plc
50 Bedford Square, London, WC1B 3DP, UK
Bloomsbury Publishing Ireland Limited,
29 Earlsfort Terrace, Dublin 2, D02 AY28, Ireland

GREEN BOOKS and the Diana logo are trademarks
of Bloomsbury Publishing Plc

First published in the United Kingdom 2025

Copyright © Jon Broome and Nick Grant, 2025

Jon Broome and Nick Grant have asserted their rights under the Copyright, Designs and Patents Act, 1988, to be identified as Authors of this work.

For legal purposes, the photo permissions on page 281 constitute an extension of this copyright page.

All rights reserved. No part of this publication may be: i) reproduced or transmitted in any form, electronic or mechanical, including photocopying, recording or by means of any information storage or retrieval system without prior permission in writing from the publishers; or ii) used or reproduced in any way for the training, development or operation of artificial intelligence (AI) technologies, including generative AI technologies. The rights holders expressly reserve this publication from the text and data mining exception as per Article 4(3) of the Digital Single Market Directive (EU) 2019/790.

Bloomsbury Publishing Plc does not have any control over, or responsibility for, any third-party websites referred to or in this book. All internet addresses given in this book were correct at the time of going to press. The author and publisher regret any inconvenience caused if addresses have changed or sites have ceased to exist, but can accept no responsibility for any such changes.

A catalogue record for this book is available from the British Library.
Library of Congress Cataloguing-in-Publication data has been applied for.

ISBN: PB: 978-1-3994-1713-6; ePub: 978-1-3994-1712-9; ePDF: 978-1-3994-1711-2

2 4 6 8 10 9 7 5 3 1

Design by Nicola Liddiard at Big Orange Door
Printed and bound in Dubai by Oriental Press

To find out more about our authors and books visit www.bloomsbury.com and sign up for our newsletters. For product safety related questions contact productsafety@bloomsbury.com.

Contents

Introduction **8**

1.0 DESIGN **10**

1.1 What is a good building? An introduction to the key principles of good buildings **12**

1.2 Beauty, utility and economy in action Archives and museums provide useful insight **22**

1.3 Buildings as sculpture When form overrides function to produce bad buildings **36**

1.4 Eco-bling – innovation and aspiration Aspiration and innovation can bring questionable results **44**

1.5 Problem-solving designs The features of a good building design process **52**

1.6 The search for quality What the designer must bring to the process – the need for care **58**

1.7 A modern vernacular Four principles coming together to shape a process for designing and producing good buildings **62**

1.8 The modern vernacular in action Applying the four principles in action in the field of housing design **72**

1.9 Self-build Lessons from an approach to housing more akin to the idea of a modern vernacular **82**

1.10 Sustainable neighbourhoods How the principles we have learnt from housing can inform good places, towns and cities **92**

2.0 REDUCING ENERGY **104**

2.1 Comfort and sufficiency Both important considerations, but what kind of 'comfort' and how much space do we need? **106**

2.2 Closing the performance gap Why many buildings do not achieve their planned minimum-emission performance 114

2.3 Form factor, massing and shape How size and shape are often fixed before detailed cost, energy or structural analysis are performed 118

2.4 Environmental modelling and targets Numbers and targets are crucial to an understanding of how to improve the performance of buildings 124

2.5 Passive solar design Shifting the focus from passive solar gains to glazing design to serve the comfort of occupants 132

2.6 Why Passivhaus? The benefits, limitations and myths surrounding the use of this highly effective building standard 146

2.7 Buildings must breathe Draught-free construction – and the resulting need for moisture and odour control – are essential 164

3.0 ENVIRONMENTAL IMPACT CONSTRUCTION 170

3.1 Heavyweight or lightweight? Advantages and disadvantages of both approaches and some myths explored 172

3.2 Upfront carbon emissions How to reduce emissions arising from the act of construction, alongside other environmental impacts 178

BELOW Brooke Lodge Farm.
A high-performance multigenerational
home by Jon Broome Architects 2012.

3.3 The more details, the more devils Principles for detailing to create robust buildings that achieve design, performance and sustainability aims **190**

3.4 Elements of good building Approaches to the design of foundations, raised ground-floor construction, non-loadbearing walls, flat roof construction and windows **198**

3.5 Services Efficient building services should be integrated into the design from an early stage **218**

4.0 THE BUSINESS OF BUILDING **226**

4.1 Land Focusing on the issue of land ownership and value, huge influencers on what is built **228**

4.2 Design education and post-occupancy evaluation How to acquire the skills, knowledge and values designers need; learning what works and what doesn't **234**

4.3 Cost and value Why do costs vary? Who benefits from value? How to reduce uncertainty and control costs **240**

4.4 Value engineering in design Understanding how buildings perform and learning from experience **244**

4.5 Appointing a design team and contractors A constructive approach to finding the right team and managing costs, quality and time **252**

4.6 Risk management and regulation Risks that cannot be avoided and how to mitigate them, and the uses and limitations of regulation **262**

Afterword **274**

Notes on the authors **277**

Chapter notes **278**

Acknowledgements **280**

Image credits **281**

Index **282**

OPPOSITE Architectural isms
come and go but good building
is timeless.

**THE MISFITS' ARCHITECTURE
CHALLENGE**

--

People wishing to 'add beauty' should be
made to prove that what they are adding:
- Does not compromise the performance
  of the building,
- Can be achieved without the use of
  additional resources, and
- Actually is beautiful

Graham McKay, Misfits' Architecture Blog

Introduction

In the UK we spend between 80 and 90% of our lives in buildings. They provide safe and comfortable places to live and work, but their construction, renovation and operation now account for about 40% of anthropogenic carbon emissions. Ironically, no building – however 'climate resilient' – can protect us from the expected increases in droughts, floods, wildfires and famine.

Many of us have long embraced the need to radically reduce the environmental impact of buildings, but in objective terms, progress has been pitifully slow. This may be in part because buildings are largely judged on their external appearance, not on their performance. In this book, we share examples of buildings in which clear performance requirements have not been met or performance has been largely disregarded at the expense of a formal concept. We also apply critical thinking to some self-evidently deep green approaches.

Throughout the book, ideas are suggested that could change attitudes to building design in order to prioritise problem-solving and the search for quality. This book advocates a people-first approach, in which occupants should take an active part in the design process. It also explores how that can extend to controlling the process in the context of self-build. If we can design with people in mind, ideally engaged as equals in the process, then perhaps better buildings – even *good* buildings – can be created.

This book discusses the idea of a modern vernacular, which makes explicit the features of the built environment that go to make delightful yet functional and economic places, and how that can create sustainable neighbourhoods. Various practical issues in low-energy design are considered, including the gap between design and actual performance; the influence of form on energy consumption, cost and upfront emissions; the need to limit construction to what is sufficient; and the limits of passive solar design. It also provides an overview of some principles of low environmental impact construction and services, with an emphasis on the need for robust detailing of the elements of a building for longevity and adaptability. We explain why we advocate the Passivhaus standard as currently the most effective approach to improve comfort while reducing energy consumption, with insights into fenestration, airtight construction and ventilation. We also highlight some shortcomings, acknowledging that the landscape changes and we have been wrong before.

The book concludes with an assessment of various aspects of the building industry and the need for reform in areas of land use, design education, post-occupancy evaluation, cost control, the design team, contracting, risk management and regulation. Through the lessons we have learnt as builders, we hope to inspire and connect with others wrestling with similar issues.

RUSS Community Land Trust, Lewisham, South London 2025. Designed with residents as a model for a sustainable neighbourhood. Strategic design by Jon Broome Architects, co-design and scheme design by Architype and detailed design by SEH Architects.

1.0 Design

1.1 What is a good building? An introduction to the key principles of good buildings

1.2 Beauty, utility and economy in action Archives and museums provide useful insight

1.3 Buildings as sculpture When form overrides function to produce bad buildings

1.4 Eco-bling - innovation and aspiration Aspiration and innovation can bring questionable results

1.5 Problem-solving designs The features of a good building design process

1.6 The search for quality What the designer must bring to the process - the need for care

1.7 A modern vernacular Four principles coming together to shape a process for designing and producing good buildings

1.8 The modern vernacular in action Applying the four principles in action in the field of housing design

1.9 Self-build Lessons from an approach to housing more akin to the idea of a modern vernacular

1.10 Sustainable neighbourhoods How the principles we have learnt from housing can inform good places, towns and cities

1.1

What is a good building?

Let's start by trying to define what makes a good building. Making buildings better has been our life's work and we are confident that we have some practical lessons to share. With so much wrong with current buildings, there is much that can be done better.

We are deliberately setting the bar at a modest 'good', very aware that the phrase 'the best is the enemy of the good' is particularly relevant in the world of building design. The final paragraph in Andrew Saint's book, *The Image of the Architect*,[1] seems as relevant today as when it was published in 1983:

> *Finally, can architecture survive as a special and unique profession if the 'imaginative' element is curbed? That is possible only if some such goal as 'sound building' in itself an uncharismatic target, can be raised to the level of ideology still enjoyed in the schools by the endless debate about styles. Otherwise, duller people will dominate and the profession will become indistinguishable from others serving the construction industry. If a generation's imagination can be fixed upon something above the game of styles, novelty of appearance, and paper projects, and remain equally resolute in the face of the allurements of commerce, we may at last get a profession worthy of the claim of leadership in that industry.*

It is easy to suggest ways to make bad buildings **better**, but it is far more challenging to find examples of good buildings. Before reading on, try and bring to mind some examples – tricky, right?

Let's try and unpick why it is so difficult to come up with objectively good buildings. There are books and websites full of buildings claiming to be great, important and iconic, but how many of these could be described as good?

Key qualities for a good building

It is relatively easy to come up with some essential qualities needed for a good building. The following list of 10 key properties is not exhaustive and most of these topics will be unpicked in some detail within the following chapters:

1. Functionality
2. Aesthetics
3. Durability and quality
4. Sustainability
5. Adaptability
6. Integration within the surroundings
7. Comfort and wellness
8. Safety and security
9. Cost-effectiveness
10. Community and social interaction

As we start to think of buildings we have been inspired by, we realise that few have all – or even most of – the qualities that we identify as making up a good building. This raises a number of questions. Are the various requirements in conflict so that any building must be a compromise? Given that many of the buildings that might be admired were built some time ago, it is unlikely that any of these will meet current expectations for comfort, health, energy use, fire safety or accessibility, although we may still make a point of travelling great distances to experience them. It is reasonable to understand expectations at the time they were built were different, but can they be brought up to date without losing what makes them special?

The old adage that you 'can make a building quickly, cheaply or well; you can choose any two' feels like a truism, but is it? Can't a good building be completed on time *and* on budget? Can't there be affordable, functional, energy-efficient buildings that people also like the look of? Is compromise inevitable? If so, who decides what is prioritised?

How do we measure good?

While factors such as cost, energy use and upfront carbon emissions can be quantified, we are still faced with apparently subjective decisions as to how low energy use needs to be. We will be discussing how to quantify the thermal efficiency of a building later, but how big a building does a person need? How much financial and material resource investment is reasonable? What is fair?

That is the objective bit, but what about beauty, character or appearance – the subject of most discussions in architecture? While it is considered shallow to judge a book by its cover or a person by their looks, it is considered acceptable to judge a building from an image in a magazine (ideally with no people present and the interior staged by a stylist, rather than ruined by the clutter of people's lives). Buildings are listed for protection based

ABOVE Hellman, architecture versus life.

on their looks, the fame of the architect and their historical significance; not on how well they perform. A building has never been listed because it is really comfortable, has low energy use or is a great use of space; indeed, many listed buildings have terrible performance. Anyone who has wanted to improve the performance of a listed building will often face opposition to making any improvements, whether that is achieving accessibility for wheelchair users or improving the comfort or energy efficiency overall.

The authors are happy to accept that there may be things that humans and even other animals recognise as beautiful to the senses. Christopher Alexander, whose 'Pattern Language' has been an inspiration to many, appears to have been searching for a fundamental truth as to what is beautiful. There is evidence that this discussion is as old as human civilisation, but beauty existed long before humans were there to appreciate it. One might wonder at the beauty of flowers but it is the insects they are trying to attract, not us. Indeed, devoid of ultraviolet vision, humans only see the surface beauty.

While interesting, we are deeply suspicious of endeavours to define beauty, although it is hard to resist passing judgement on matters of 'taste'. The problem for us, in the context of the elusive good building, is when a rigid idea about what something 'should' look like comes at a cost to another aspect of the building.

Vitruvius, the first-century BC Roman engineer, artilleryman and architect, is attributed with the snappy *firmitas*, *utilitas* and *venustas* as requirements for a good building (this usually gets translated as strength, utility and beauty). Strength can more usefully be interpreted as durability rather than simply mechanical strength, resisting gravity or siege engines; these days it is more usually moisture that does the damage.

This seems like a good starting point. Clearly our requirements have changed a bit in the last couple of thousand years, but most of our modern requirements can fit into these three categories. What is missing is economy. Perhaps in the days of abundant raw materials this was not so important? It turns out that while he didn't make it one of his big three, Vitruvius had something to say on economy that seems remarkably up to date:[2]

Economy denotes the proper management of materials and of site, as well as a thrifty balancing of cost and common sense in the construction of works. This will be observed if, in the first place, the architect does not demand things which cannot be found or made ready without great expense.

It might be worth adding affordability or economy to this high-level list because, for most people, this is important. In the interest of simplicity, durability can be included as part of utility so we have: beauty, economy and utility.

Of course, all three categories overlap. We can argue that an ugly building will be demolished regardless of how physically durable it is, or that a building that is 'iconic' achieves the function of putting a city on the map - the so-called Bilbao Effect.

 Inspired by the form of diamonds at the molecular level, the Sheffield University Diamond Building replaced the Edwardian wing of the Jessop Hospital.

Beauty

Beauty is not just visual here. We include all the senses but also history, rarity, character, fame, wow factor, quirkiness, kerb appeal, soul, style, trendiness, homeliness, feelings of comfort, *hygge*, atmosphere – anything that is not to do with physical function.

As with the other categories, a building can't be judged in isolation and out of context. We are happy for the qualities in this section to be subjective, but they can be discussed. We can argue whether a building should make us feel comfortable or challenged without reaching consensus, but we should be able to get a reasonable agreement as to whether a given building makes most people feel comfortable or challenged. These are the qualities that a lot of architecture focusses on, but in our search for what makes good buildings we are mostly interested in when these ideas are in conflict with utility and affordability.

Economy

Economy is about cost and affordability. It is about cost relative to a budget or other constraints. It is not about value; that is relative to utility and beauty. If costs can be

RIGHT Variety in the high street - some may be good buildings.

reduced without reducing beauty and utility then we are increasing value, whether it is so-called 'affordable housing' or a millionaire's mansion. As with beauty, any absolutes are not inherent in the category. If a government is looking to decarbonise a country by a certain date, that will require a budget.

Economy also relates to available materials and human resources. Again, the context and time in history will have a large impact on cost. What was a cheap mud and straw shelter for a family of agricultural workers is now a Grade II-listed desirable residence with rustic charm. What was built as speculative housing for the wealthy can become derelict until the area is eventually gentrified over 100 years later. Affordability can also include social concerns, which will include who the client is.

Utility

Utility includes durability and general fitness for purpose. As a minimum, most buildings have always been required to provide basic shelter and security but modern standards of comfort, health and security are higher than in the past. We would include all functional and performance requirements of a building, including the energy use, fire safety and accessibility. Again, there are no absolutes as standards rise and fall over time and various buildings have very different functional requirements. We can judge a bus shelter or a hospital as a good building, but it makes no sense to compare them with each other.

How to describe 'good'

We are trying to find a simple map that helps us set a brief for a good building or to discuss how well any building rates once it has been built. A radar diagram of every imaginable quality could be used to rate any building against key performance indicators. This is a useful way to judge performance of a specific building against a whole raft of performance targets – from energy use to space standards to occupant satisfaction – but the importance of each of many metrics would need to be assessed. Even with weighting, such a diagram will be skewed as decisions will have to be made about what is or isn't included from an effectively limitless list of possible qualities.

There is a need for a much simpler higher-level model so buildings can be discussed in the round without having to constantly deal with 'not just aboutism'. An example of this is when a member of the design team suggests a change to reduce energy use and someone responds that it is not just about energy, effectively stopping any attempt at integrated design. It is only fire and structural engineers who are largely immune from this judgement. No one would say 'it is not just about the building not killing people' – at least not out loud, in the wrong company.

We want to try and answer the original question about what makes a building good, but more interestingly, whether any failing example can be made into a good building at the design stage or as a retrofit. Or are some qualities that make a good building mutually exclusive, so there will always be a choice? Arguably the main driver for us writing this book is that we believe it is possible!

ABOVE We need to question everything but we also need to identify and embrace what works and is good enough.

Key properties

Across their work, Charles and Ray Eames suggested that the secret of good design is to identify and embrace constraints. If key constraints can be identified in each of our three categories, we might have the chance of creating a good building.

Having reduced our categories to three – beauty, economy and utility – we can represent our families of qualities as a Venn diagram.

Aesthetics

We can place buildings we know on this map. If we start with one of the many lists of *best* or *iconic* buildings, we find that most are good-looking – beautiful even, but with little or no overlap into 'function' or 'economy'. We are not concerned here about which of these buildings are more or less beautiful, but rather on how useful or economic they are.

Few will be cheap, but more relevant, most will have cost significantly more than the original budget. As well as budget overruns, we will also find stories of woefully poor performance, including roofs that leak and houses that are too cold in winter and too hot in summer. The 'Farnsworth House' by Mies van der Rohe or 'Fallingwater' by Frank Lloyd Wright are well-documented examples, yet both are highly acclaimed as iconic and innovative architecture.

What is more, sometimes it seems that the more famous and influential the building, the higher the cost and lower the utility. Please try and think of exceptions to this – we are genuinely looking for any examples of good buildings that sit in the middle of the diagram.

Some may argue that a great building does not need to be functional or economic and that celebrity is enough. We disagree; we are not arguing for the end of celebrity, we just don't see more celebrities, or even the celebration of celebrity – whether of people or buildings – as being the answer to any of the world's problems.

Economy

If we did a redesign, could we make any of these icons cheaper and better performing, perhaps moving them to that elusive centre of the Venn diagram? Modern materials may allow us to waterproof flat roofs, and some buildings could be greatly improved by adding insulation and modern glazing, but we will often struggle to improve the comfort and thermal performance without destroying the very things that put many of these buildings on a pedestal. Direct glazed, thin steel window frames and concrete cantilevers leave no room for insulation, so the look is going to have to change – and probably not for the better.

We are not picking on modernism here – listed historic buildings could be placed in this category. Thanks to more constraints in former times, many of these buildings will actually perform better than more recent icons. However, if we are to upgrade the utility by insulating – adding secondary glazing, a heat pump or a wet room – we may face strong opposition as all these measures may be seen to compromise the 'beauty' of the building in unacceptable ways; in the UK, we may even be committing an imprisonable offence.

We can imagine cheap buildings that are not particularly beautiful and are of poor

ABOVE Houses designed to achieve the highest performance within a normal budget for social housing. Jon Broome Architects 2005.

utility; no surprises here except that – as with the historic buildings – the utility may well be better than our expensive, iconic examples. This is no coincidence, as cheaper buildings will be forced to use a simple form with proven standard details. For example, tiled truss rafter roofs tend not to leak or rot and are easy to fix.

With some thought, examples can be found that are both economical and beautiful. Here we find a lot of vernacular buildings that were very economical when they were originally built. All will have proved their durability by still existing and some may now perform surprisingly well in terms of comfort and shelter, particularly summer comfort. Many of these buildings will have originally been smoke-filled, rat-infested hovels made from mud, sticks and agricultural waste such as cow dung and straw. Housing farm labourers at minimum cost and close to their place of work were the main considerations. That such buildings became the subject matter of romantic paintings by the masters – and more recently, extremely desirable and expensive residences – was never the original intent.

At some point in the recent past, these buildings might have sat in the elusive middle of the Venn diagram. Mains electricity, piped water, indoor toilets and central heating powered by cheap energy were all embraced. But the energy crisis and climate emergency have moved the goalposts for acceptable performance for a good building.

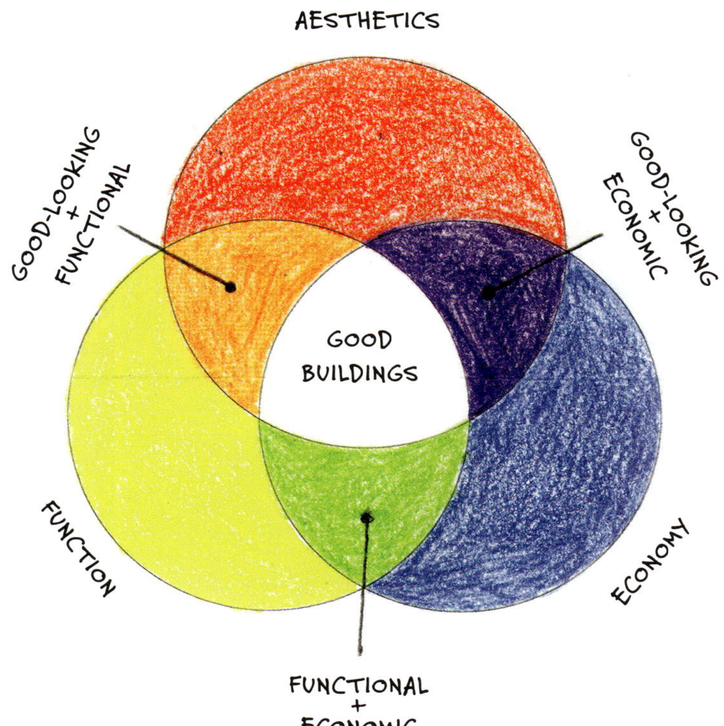

LEFT Are function, aesthetics and economy really mutually exclusive?

There were some interesting attempts to deliver beautiful economic buildings as a utopian ideal, but most performed badly. Look at the Hunstanton School, designed by Alison and Peter Smithson in 1954, for instance. This Grade II listed glass and steel building was highly influential on school design and was described by architecture critics as the New Brutalism. Locals called it the greenhouse because the glass and steel envelope was hot in summer and cold in winter. If you can picture an example in this category, could it be upgraded to modern standards of energy performance without losing its 'beauty'?

Function

Finally, we have highly functional buildings. Interestingly, if we describe a building as functional, it has immediate negative visual connotations and implies a certain Soviet era aesthetic. To say a building is very functional carries some baggage that is not there if we say a building works really well – arguably the same thing. We imagine some sort of low-cost, ugly but functional building, and it is interesting to think what could be done to make it beautiful. Would this make it expensive or reduce its function?

It is usually easier to make a cheap, functional building beautiful than it is to make a beautiful building functional. This can be applied to any building you might like to think of. After all, a can of spray paint and a card template can turn an underpass into an art gallery. Perhaps it is in the middle of the Venn diagram that we find form following function, not as some sort of style or design theory but as a solution to the problem of creating a good building.

Where buildings are designed based on abstract visual concepts, there will, in our experience, be huge challenges in getting the building to function well. The next chapter explores an example of a mining museum inspired by a coal-cutting machine, and there are a surprising number of libraries in the shape of books. Perhaps a building can be conceptual, functional or cost-effective – pick any two but you can't have all three. But beautiful, functional and economical *is* achievable, even if this will mean different things to different people. This should be the aim.

It is nice when beautiful form follows from functional constraints. However, that isn't a law of nature, although, thanks to selection pressures, nature seems to be brimming with examples. See Chapter 1.4 for a critique of designing from nature. What is self-evident is that what a person might think is beautiful may not be functional, and sometimes that is the point. Low function has always been a signal for wealth, power and status.

Robert Pirsig, author of *Zen and the Art of Motorcycle Maintenance*[3] spent his life exploring ideas around quality but frustrated his readers by never defining it. In a 1974 lecture following the publication of his first book, he said 'As long as you keep it undefined, then it becomes an instrument of change, and you can grow, because the things that you find Quality in are going to change as you grow.'[4]

OPPOSITE Hunstanton school, 1954, listed Grade II*. Elegant and economic steel and glass 'brutalist' building - but many of the single glazed steel windows have now been replaced with timber. Note too, the addition of blinds.

Beauty, utility and economy in action

Archives and museums are a niche building type, but they provide useful insight and evidence of some of the issues we want to raise. They have always had very clear environmental requirements for temperature, humidity and to protect the – often priceless – contents from decay, fire, water and theft. These requirements allow us to objectively judge if the design has worked or failed in a way that we cannot, say, for a pavilion or even a house in which the occupant might be willing to suffer poor comfort and high bills in order to live in a piece of architecture.

Charles Eames referred to this as a 'kite problem', saying:[1]

> The marvellous part about a kite problem is that this is one area in which one can definitely judge its success or failure – that it will fly or it will not fly. I wish more problems could be so beautifully defined.

Ray and Charles Eames loved kites and their focus was on fun and beauty. They were not building kites for heavy lifting or altitude records, but if a kite does not fly its beauty is irrelevant.

Archive and museum buildings are a kite problem. The bar for aesthetics is often higher than normal, but they must have utility and economy. This book will explore some examples of the results of two approaches to design: architectural inspiration by some of the most revered architects in the land and design by problem-solving (*i.e.* what is just called 'design' outside the worlds of architecture and fashion). This is the approach we are trying to make a case for throughout this book.

While the technical requirements for an archive, collection store or museum should be clear and non-negotiable, there is usually also a requirement for the building to win over donors, funders or a competition selection panel who will be judging concepts and visual images. At some point, the reality of function and available budget will need to be addressed if it is to be built.

As a minimum requirement, our base case building needs to address beauty, utility and economy; there is no room for compromise in such a building.

- Beauty – this is usually important to attract funding, and visitors in the case of a museum. As a minimum, it must meet any planning requirements.
- Utility – there are clearly recognised standards for museum and object stores covering temperature, humidity, fire, security and safety requirements.
- Economy – funding is usually with public money. The budget is typically fixed, whether large or small. It is easier to raise capital for exciting new buildings than for ongoing running costs and maintenance, but this must be considered.

Inspired archives and museums

In the previous chapter, we made the far-from-original claim that often the most revered architecture has the lowest function and highest cost. The following small sample of well-known buildings seem to prove this case.

What is surprising is that where famous architects have designed buildings that fail to meet the implicit and explicit brief to protect the collections, the architecture has been valued more highly than the collections themselves. The proof of this is that any suggestions to change the architecture to try and solve the problems it has created are met with strong resistance.

In the less than unique examples overleaf, it is clear that the architectural concepts – rather than the build quality, lack of budget or poor mechanical and electrical (M&E) services – are the reason for the failures. While we see a strong correlation between famous architects and poor performance, this is not a causal link.

The first three examples and their designers have all received critical acclaim and are based on abstract concepts. All perform badly because of the very concepts that make them iconic. For the sake of this discussion, we are happy to assume they all meet the highest standards of beauty (that is not a discussion we will get into here, but the reader will have their own opinion on this subjective aspect of the designs).

The latter three examples are primarily designed to maximise utility and protect the collection. Considerable effort went into client consultation working with archivists, conservators and other building users. Consideration was given to fire, security and environmental stability, as well as future flexibility and user preference for collection access. All minimise reliance on energy-consuming mechanical plant by utilising proven scientific principles of passive conservation.

Whether they are more or less beautiful than the other examples is a matter of opinion, but they showcase how design can be used for problem-solving. All three were considered good enough by planners and clients, and two have won architectural awards.

The Ruskin Library, Lancaster

MJP Architects, 1997

The Ruskin Library has won multiple awards, including the *Independent on Sunday* 'Building of the Year'. It has also received critical acclaim from the Design Council, Royal Institute of British Architects (RIBA) and the Civic Trust.

The architect boldly claims that this building reimagines how archives are designed. In our opinions, the body of knowledge on conservation physics and low-energy archive design embraced by Tim Padfield and Poul Jensen in their 1990 article[2] is ignored in favour of metaphor. Apparently, it is a 'gondola' because Ruskin liked Venice. Long grasses were to surround the building, giving the illusion of water moving when the wind blew. Mundane practicalities of meadow management and the fire risk of tall, dry grasses surrounding an archive means the building is now surrounded by a neatly mown lawn.

The inspired concept was to house the unique archive in the centre of the building surrounded by office, gallery and circulation space to isolate the collection from the assumed hostile outside environment, without the need for air conditioning.

Unfortunately, this ignores the most basic physics and the second law of thermodynamics. This means the archive will always be warmer (thanks to gains) than the surrounding heated and habitable space, and so it will need cooling all year if temperatures are to be kept low enough to satisfy conservators. Cooling was not included in the design for the Ruskin as the architectural inspiration and thermal mass was to provide the ideal conditions.

Although traditional security may be found wanting, metaphorical security was provided by the glass-floored corridor that surrounds the central strong rooms representing a moat to protect the collection. This has been obscured, perhaps following complaints by skirt wearers!

There seems to be no information in the public domain on the environmental performance to back up the bold claims made for the design, but we do know that the building has been closed from 2021 to 2025 for major refurbishment.

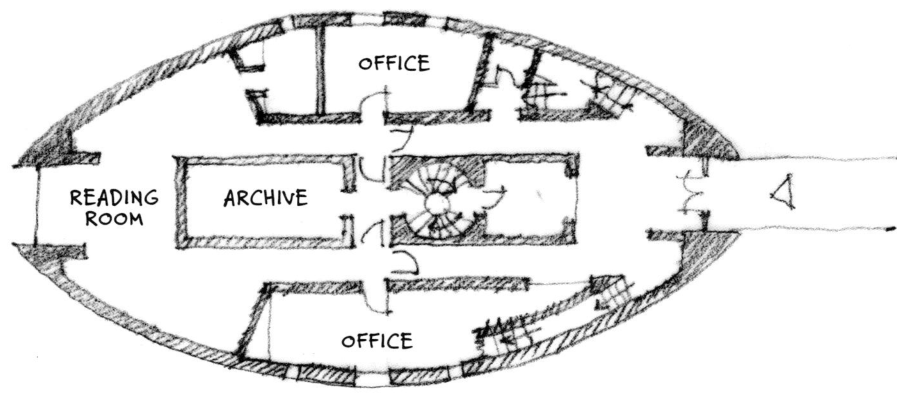

American Air Museum, Duxford

Foster + Partners, 1987–1997

The creative genius of this building is obvious from Norman Foster's fluid and impressionistic conceptual sketches created in a moment of inspiration, although it took another 10 years for the engineers to convert the sketches to concrete. The result has been called one of the finest pieces of modern architecture ever built, now listed as Grade II – a rare accolade for such a new building. It has received more than 10 prestigious awards, including the Stirling Prize.

The building has been likened to a Second World War blister hangar, according to Foster + Partners' website. It is deeply ironic that the cheap, elegant and effective blister hangar – a good example of design intelligence – only provides the stylistic inspiration for the shape of one elevation of this building.

Despite all its accolades, the building fails to meet the minimum requirements for a museum displaying historic artefacts, or for the health and comfort of staff and visitors. Admirers will point to the connection to outside, the rhythm of the panels and glazing bars, the massing, the ironic references and so on. In objective terms, the surfaces of the historic objects in summer can be too hot to touch and the air temperature can exceed 30°C. The light level from the large area of single glazing is causing visible damage to exhibits such as the rubber tyres. The extreme glare makes it difficult to view the objects on display and almost impossible to see the contents of the dimly lit glass cases that contain the more sensitive objects.

Some have said that the requirements for environmental conditions were different at the time the building was designed, and that is true; they were much tighter and have been relaxed in recent years to allow slow seasonal variation in the interests of energy efficiency. Although inspired by the shape of a Second World War hangar, the building lacks hangar doors. Removal of objects that won't fit through the pedestrian doors requires a crane to dismantle the glazed façade.

A simple fix to the glare and overheating would be to add external shading to reduce solar gain and light levels in general. This would protect the contents from heat and light and greatly improve the experience for staff and visitors, allowing them to view objects as more than silhouettes. However, this would change the award-winning look and would require an admission that the design was fundamentally flawed.

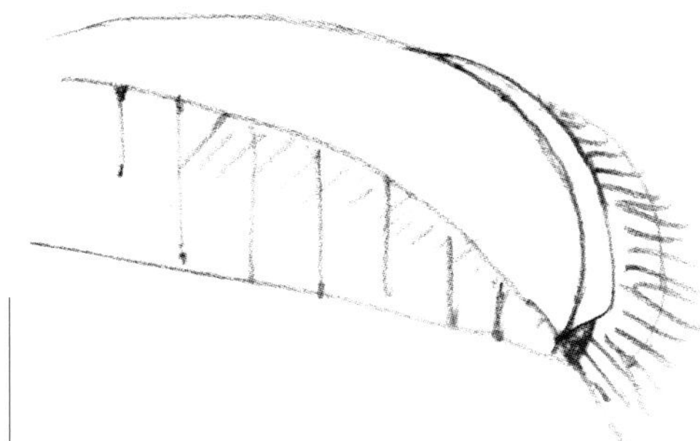

ABOVE Jon's parody of the original inspirational sketches for 'one of the greatest pieces of modern architecture ever built'.

ABOVE The Cutter with signage
for anyone who might
otherwise miss the reference.

RIGHT The coal cutter inspired
form creates problems that are
not resolved.

Woodhorn Mining Museum and County Archives, Ashington
Tony Kettle, 2006

When redeveloping the Cutter, Tony Kettle was inspired by monster coal cutting machines. From the ground, this building can easily be dismissed as a motorway service station with gabions, but an aerial view reveals the concept. As with the other examples, the museum contents and staff more than compensate for any shortfall in the building, as do the preserved colliery buildings on the site of this former coal mine.

While the convoluted and highly glazed museum is challenging to heat and cool, the actual archives are simple windowless concrete boxes constrained by the need to fit roller racking, which should work well. They are airtight, insulated and thermally coupled to the ground, and would be capable of maintaining excellent environmental conditions with the addition of about 1kW of dehumidification for all five archive stores. This would make it one of the most energy-efficient and stable archive buildings in the world – in spite of the inspirational design, not because of it.

Unfortunately, fitting simple boxes into the Cutter-inspired plan leaves awkward, deep triangular rooms, which are grim workspaces with lots of awkward sections of flat roof that seem difficult to keep watertight. The triangles contain heated offices, and in one case a boiler room, so the heat gains are considerable, and archive temperatures get higher than desired now the very expensive air conditioning has reached end of life.

There are lots of other examples of similarly inspirational archives and museums that either fail to perform or require considerable energy input to achieve the required conditions. It may be unfair just to pick on the three examples here, but the themes tend to repeat.

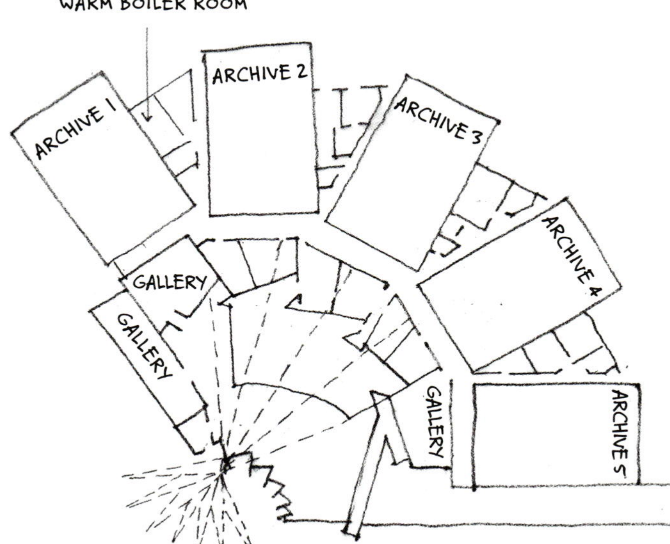

Intelligent design

Suffolk Records Office, Ipswich

Henk Pieksma, 1990

This records office shows a rational design driven by thermal and fire performance with a robust pitched roof and simple cavity construction. Hygrothermal modelling was carried out from an early stage.

Designed with the help of conservation physicist the late Tim Padfield, the building performed very well with no air conditioning. The average annual temperatures remained below 18°C but the summer temperature rose a little over 20°C so air conditioning was unnecessarily added by the client. More recent standards embrace seasonal drift as long as average temperatures remain under 18°C. The slightly higher summer temperature helps maintain a stable relative humidity level without the need for dehumidification.

The winning of architectural awards was not part of the brief or of interest to the designers. The building has been ignored by social media influencers and the architectural press but, unlike the highly acclaimed examples above, it works.

ABOVE Economic build with good performance and very low energy use.

2004-2006

LEFT The graph shows temperature drifting with the seasons which keeps the relative humidity (RH) within a very acceptable 50-60% all year. The freely published lessons from this and other buildings have made the following buildings possible.

Hereford Archive and Records Centre

Architype, 2015

The council needed to replace a failing archive building but the available budget was less than would normally be required for a conventional archive. The design team set the brief for a very low-energy building, inspired by the conservation physics work by Tim Padfield and others.

The form followed from many practical considerations and constraints including:

- simple thermal separation of archive and offices;
- plan and section evolved from racking and storage requirements and staff workflow;
- consideration of security and fire from an early stage;
- consideration of future expansion at an early design stage;
- orientation considered daylight and solar control with the repository to the north;
- workflow between the repositories and the public reading room;
- a tight budget.

The resulting massive archive box was then considered visually. Cedar shingles were chosen to provide texture, and we think it looks good. Interestingly, Hereford Archive and Records Centre does have an actual rather than metaphorical moat between the road and the building. This is part of the security strategy against ram-raiding and also acts as a swale for the surface water drainage.

The building has won a small number of awards but, like the Suffolk Records Office, has not received much adulation from the architectural press and community. It does, however, receive a constant stream of visitors from other archives and is loved by staff and visitors.

OPPOSITE Physics over metaphor, HARC was informed by the freely published technical information on Danish low-energy storage by Tim Padfield and his colleagues.

LEFT Graph showing seasonal temperature drift and stable RH (four years).

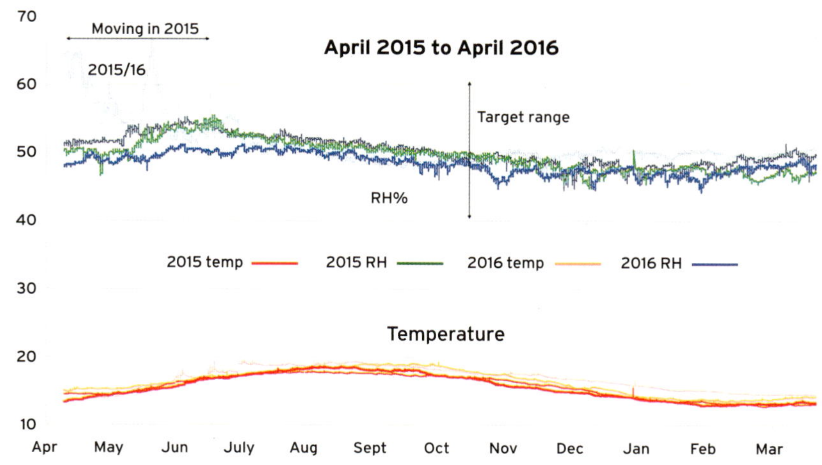

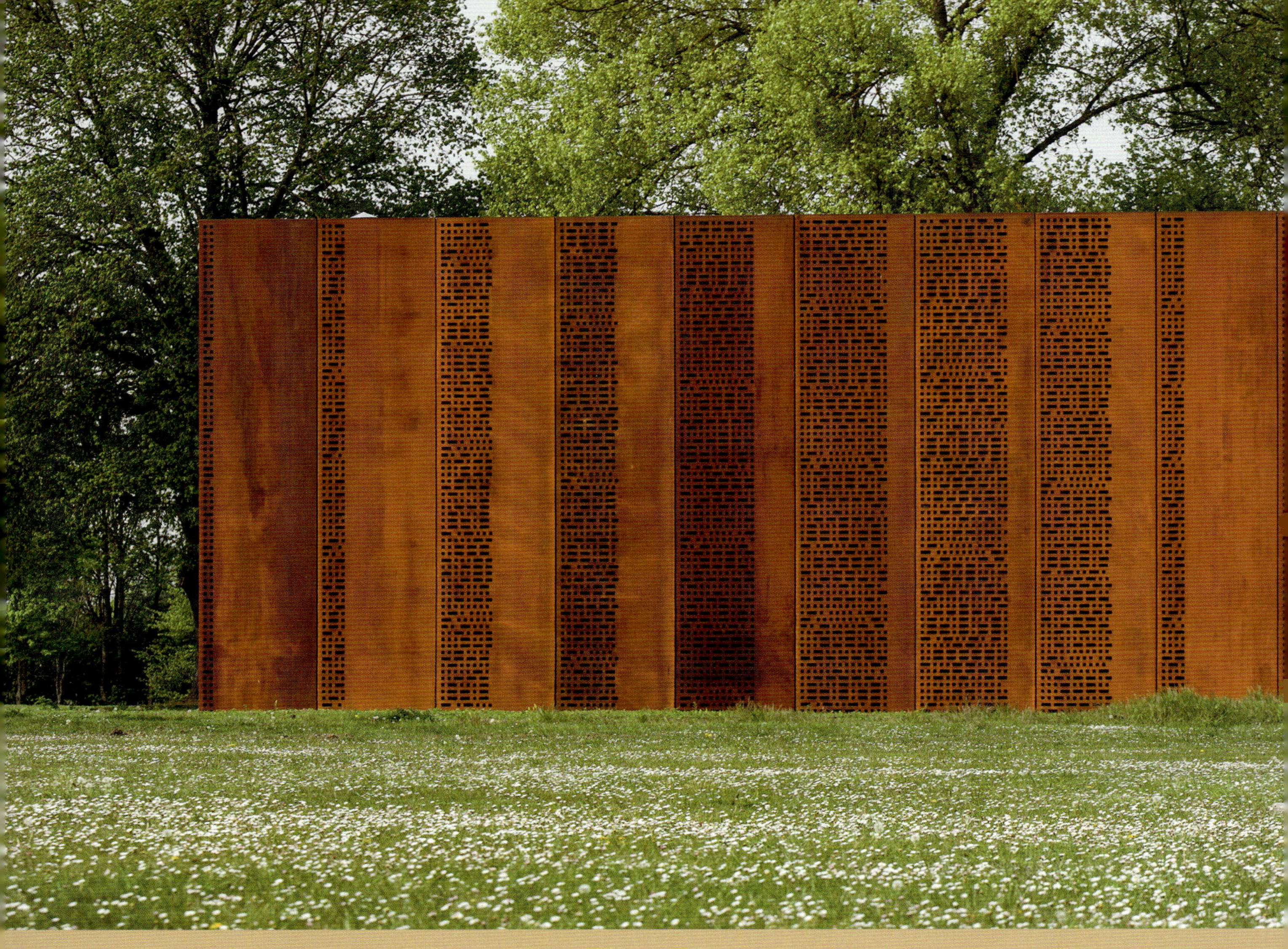

ABOVE Very low-energy
paper store.

Imperial War Museum, Paper Store, Duxford
Architype, 2019

This design followed from constraints and built on the experience and confidence provided by the Hereford project. Architectural inspiration was added later, due in part to client expectations. Once it was clear that the best solution was a simple box, the 'boxiness' was fully embraced and all rogue articulation, such as doors, entrance canopies, rainwater downpipes and pressure relief vents for the fire suppression system, were carefully hidden rather than expressed. All this added cost but only had a relatively small impact on the building performance.

The plan proportions were driven by maximum roller racking lengths (building width), precast roof plank spans (bay width), preferred racking height (section) and total amount of storage (building length).

The external loading bay was changed to an inset one to maintain the pure box form. This added cost and complexity - aiming to look simple often does this.

The building has won a number of architectural awards, including three from the RIBA, although these were probably because it is a Corten box rather than because of the exemplary design, which is largely invisible to the untrained eye.

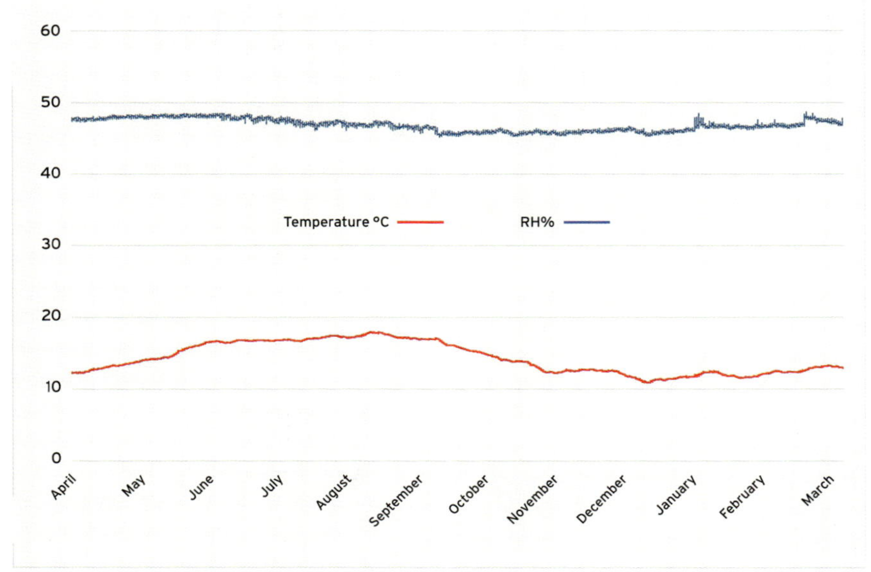

LEFT Graph showing seasonal temperature drift and stable RH without heating or cooling.

Somerset Heritage Centre, Taunton

Roberts Limbrick, 2009

While lack of starchitecture budgets normally provide some creative restraint, and so better-performing buildings, we find that even quite run-of-the-mill buildings can be made worse by the smallest amount of architectural inspiration.

The Somerset Heritage Centre is at first sight a modest building that does not look out of place with its out-of-town retail shed, cash-and-carry neighbour. The archive is a simple reinforced concrete box on two floors providing inherent airtightness, high security and the necessary four-hour fire protection.

The building actually works very well for the intended purpose. This is largely due to the considerable input from staff during the design process. The structure is elegant and avoids cantilevers and other challenges, but the external insulated panel façade steps out at first-floor level, leaving a large void between the insulated cladding and the simple concrete box.

This medieval jetty reference adds visual delight, as clearly seen in the photo, but makes the building worse in every objective and measurable way. A substantial steel frame holds the cladding away from the main structure and allows free movement of wind between the insulated cladding and the archive, adding cost and upfront emissions. The cost and utility of this building has been compromised by the effort to lend 'expression' to its façade.

ABOVE This thermal image taken inside the building shows the cooler (blue) wall where the external façade steps out creating a cold void behind the insulation.

RIGHT The stepped-out cladding increased cost and reduced function, does it add beauty?

Archives without designers

There is another category that is also interesting: buildings never intended as archives that have had the windows sealed up and racking installed to be used as archives; vernacular archives, perhaps.

As long as the roofs don't leak, most of these buildings tend to perform much better than higher-budget, award-winning inspirational architecture. Examples include numerous repurposed Second World War buildings and even a stone church in Edinburgh, which is performing very well since the air conditioning was switched off some years ago.

Obviously, any ancient artefact can only have spent a very short period of its life in a modern facility; sadly, when inspired architecture is involved, those might be its last years.

This chapter is intended to throw light on approaches to building design in general, but if you are interested in archives and museums in particular the writings of Tim Padfield (see page 278) are highly recommended.

We draw attention to these salutary tales to highlight the need to balance conceptual thinking with rational analysis. As Richard Feynman said, 'For a successful technology, reality must take precedence over public relations, for nature cannot be fooled.'[3]

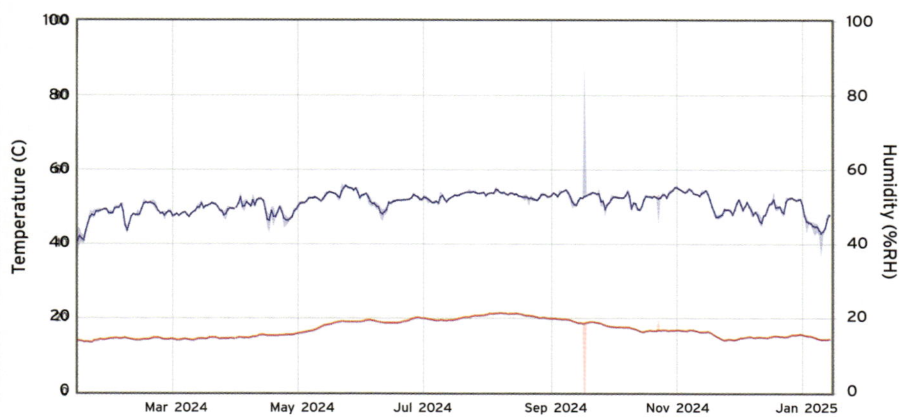

ABOVE Graph showing a natural seasonal temperature drift that keeps the relative humidity (RH) comfortably in the desired range without any plant. Daily fluctuations are far less than the annual scale suggests.

TOP A 200kW chiller being removed to allow roof repairs and never to be replaced.

ABOVE West Register House in Edinburgh is an archive housed in a converted church. With the HVAC turned off it now works better than many purpose-built modern archives.

1.3

Buildings as sculpture

We are living in a time when often the appearance of buildings and how they impact on our visual senses matter more than how effective they are in answering the purpose they are built to fulfil: places to live, work and play that are comfortable, economic and sustainable. Lavish care is taken over statement architecture whereas public housing is neglected and badly maintained, and this is a measure of the contempt of some over the many.

Post-rationalised architecture

The office of the late Zaha Hadid, another much-admired architect, devises offices, factories, museums and other buildings using designs incorporating mathematical rules and algorithms to determine shape and form. They also market expensive 'lifestyle' designs using the same software to create similar shapes for office blocks or perfume bottles. The link between architectural shape and the meaning connected to it has always been tenuous, so that 'classical' forms of columns, pediments and so on have been taken to signify, variously, Greek democracy, Renaissance harmony and Stalinist totalitarianism.

Less extreme examples abound, driven by changing fashion – the idea of 'expressing' the structure on the outside. Crown Hall at the Illinois Institute of Technology by Mies van der Rohe is a famous example, or Richard Rogers' preoccupation with showing the building services on the exterior as at the Lloyd's Building in London. This creates various significant problems with weatherproofing, thermal bridging and just mounting complex machinery in a hostile environment.

More mundane is the cliché of 'expressing' the vertical circulation, *i.e.* stairs, as a fully glazed enclosure in contrast to the rest of the building, which sensibly places dwellings or offices behind walls with windows for daylight, sunlight, view and ventilation. The 'articulation' of the façade leads to unnecessary overheating and maintenance problems.

RIGHT The fully glazed stairwell is a very common cliché and tends to increase cost and make the building perform much worse. You don't need that much glass and the space is too hot in summer and too cold in winter and it is more difficult to keep the rain out.

Architecture is treated as sculpture, with the architect a decorator dealing with the appearance of the skin of the building. This post-rationalised architecture is expensive and unnecessary. It occupies the space vacated by modernism, the approach based on a utopian outlook going back to the mid-nineteenth century that aimed to improve the world for all, employing the ethics of honesty and quality to eradicate ignorance, sickness and poverty. Design would create comfortable homes in liveable cities.

Since the 1960s and 70s, modernist thinking has been overtaken by global capital and a shift from public endeavour to private enterprise. It has been replaced by postmodern uncertainty, with an absence of any unifying political beliefs leading to a multiplicity of styles, symbols and metaphors competing for attention in the free market. Anything goes in the entrepreneurial city where power and money have co-opted the professionals.

Constructing buildings requires land, materials and money, and is a complex and costly process. Nevertheless, corporations and oligarchs still feel the need to put their mark on the world and build outstanding, expensive and sometimes slightly mad constructions. If you follow the power and money to Abu Dhabi or Shanghai, you will see that we live in an age of weird dystopian, unliveable and incidentally completely unsustainable cities.

This recalls a building designed by one of the world's most lauded architects, Daniel Libeskind. He was commissioned by London Metropolitan University to design a modest building for seminar rooms, a café, associated circulation spaces and storage on the Holloway Road in North London.

The university went to Libeskind because they wanted a building that would stand out from the humdrum environment of a busy road in North London. What they got was three colliding jagged forms, which are arbitrary and deliberately out of context with the surroundings. Inside, there are intersecting lines in the ceilings, which are confusing and lead the eye in random directions. The architect's response to the site alongside a road choked with traffic and next to a tower block, as reported by Tom Dyckhoff in *The Age of Spectacle*,[1] was to search for a hidden theme to inform the design.

TOP Apparently, the ceiling lights are aligned with the motion of the spheres, although it is unclear how. There is no link between form and purpose.

BOTTOM Graduate Centre for London Metropolitan University by Daniel Libeskind 2004. Arbitrary and deliberately out of context.

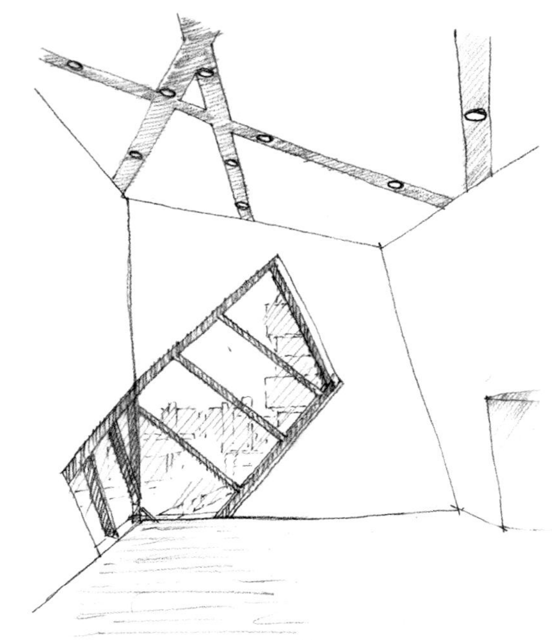

A reaction to post-rationalist building

In 2016, in a bout of aesthetic protectionism, the Chinese State Council issued a directive banning 'bizarre, oversized, xenocentric, weird' architecture. By this, they meant the kind of buildings that had been increasingly built since economic liberalisation began in the country in the 1990s – including the Guangzhou Circle, the world's tallest circle-shaped building at 138 metres high. The design of the 33-storey office building is apparently based on oriental psychology, perception and the Chinese use of symbols in writing. Many other meanings are linked with the building: the iconic value of jade discs and numerological tradition of feng shui. The reflection in the water also corresponds to the number eight, which has a strong positive value in China. The building takes reference from the Italian Renaissance – *quadratura del cerchio* (squaring the circle).

Such architecture is now thought to clash with Chinese traditions. New architecture, the Council's decree stated, must be 'suitable, economic, green and pleasing to the eye'. This attempt to legislate for sensible building is probably unlikely to succeed, even in the Chinese economy, subject as it is to close direction by the government.

ABOVE Sculptures make poor buildings and buildings usually make poor sculpture but sculpture can transform a building. Playful sculpture by Lucy Casson, Devonport Dental Hospital.

LEFT Guangzhou Circle, the world's tallest circle-shaped building.

Community building

In the United Kingdom during the 1970s and 80s a reawakening of a more progressive, co-operative and communal view occurred with a community architect in charge of RIBA. Attention was directed towards neighbourhood improvement projects, community self-build and participatory design. This thread of ideas has struggled to survive into the twenty-first century, where competition is valued above consensus and global finance and global markets rule in many areas, including housing. Nevertheless, there has been a reawakening of interest in a collaborative rather than competitive approach based on involvement of building users and residents as well as designers, builders, financiers and policy makers to supersede the free-for-all of the free market. We are still some way off from having a coherent alternative narrative that also includes alternative political and economic models to replace existing postmodern confusion.

We have had a Right to Buy since 1980 – but there is no right to a decent, affordable home. The government introduced a Right to Build in 2016 in an attempt to open up the housing market to small developers and individual self-builders with the aim of encouraging big housing developers to offer more choice, better value and superior quality. The Right to Build requires local authorities to provide sufficient planning permissions for serviced plots to satisfy local demand, as indicated by a register of individuals and groups. It remains to be seen to what extent this measure will enable a change in the housing market.

Good buildings require a balance between performance, economy and appearance. This is likely to be promoted by ensuring that residents and building users have an effective role in developing the built environment. We need architecture that engages with people – buildings designed by those who not only understand how different structures are made but also how they relate to political and financial circumstances.

Eco-bling innovation and aspiration

The late Howard Liddell coined the term 'eco-minimalism' as an antidote to eco-clichés and eco-bling.[1] He drew up a list of building technologies that were accepted as green without applying critical thinking. Our list is a little different, but we are glad to be challenged.

As Charles Eames said, 'Innovate as a last resort. More horrors are done in the name of innovation than any other.'[2] Ironically, architect Lloyd Alter has stated that he has a pile of Eames chairs in his garage because the innovative way of fixing the legs to the seat with rubber pucks failed! In solving a problem, often new ideas, methods or technologies emerge. If the new stands the test of time, it might be seen in hindsight as an innovation. To seek innovation for innovation's sake or to assume that innovation represents progress is flawed and unsustainable.

Some environmental rating schemes, planning permissions and architectural competitions reward or demand innovation, assuming that innovation in itself is a good thing, even if there is no obvious problem to solve. We all feel the pressure to come up with something new rather than repeating what we did before, however well it worked. In fashion, innovation is what drives the market; without constant innovation there is no fashion industry. In architecture, as in fashion, this innovation must be very visible.

When the word 'aspirational' is used to describe sustainable aspects of a project, alarm bells should ring. Sustainability must by definition become normal, routine, mundane even – anything but aspirational – to avoid the market pushing beyond the optimal solution. The most disappointing is the high-profile, sustainability award-winning, aspirational project.

Cork House

The multiple sustainability award-winning Cork House is an interesting case study, given that from all the hype it seems to be about as sustainable as you can get. The architect's claims are backed up by numerous consultants' reports and a life cycle carbon analysis that says it is carbon neutral.

Cork really is a remarkable material and apparently the cork industry is in decline because of a move to cheaper alternatives such as synthetic cork and screw tops. So, why are we so down about it? The architecture critic Catherine Slessor waxed lyrical about the sustainable credentials of the project in a glowing review in the Observer.[2]

Cost

Whether it is food, clothing or housing, there is a case for sustainable quality production costing a little more. Cheapest may not be best, but sustainable food, clothing or building cannot, by definition, only be for the rich. The fact that we can't find any mention of a budget price for Cork House suggests that the cost is hard to justify, even as a prototype. We can, however, do some rough numbers.

As the building won the Stephen Lawrence Prize it should have a less than £1m build cost, as that is stated in the rules. On a very modestly sized building this would equate to <£22,700/m^2 or about six to eleven times the build cost for a one-off house at the time. To try and get a better estimate, we know it is said to contain 1,268 blocks of cork, each weighing about 13kg. So that is roughly 16.5t of cork, or about 106m^3. Simple rectangular cork insulation boards cost about £1,500/m^3 at the time of writing (£3,600/m^2) for the raw material for the walls and roof, without computer numerical control (CNC) robotic machining into Lego-like blocks, sealing gaskets or site labour.

BELOW The Cork House by Matthew Barnett Howard and Oliver Wilton 2019.

The upper estimate of £22,700/m² doesn't seem so silly once we add in the floor, foundations, fenestration, joinery, services, kitchen and bathroom, sprinkler system (cork burns) and some labour and profit, but excluding the inevitable and expensive prototyping costs for a one-off – and clearly innovative – building such as this.

Potential to scale

Even if cost were to be ignored, can we grow enough cork to use it as a carbon-neutral building material in this way, at any sort of scale?

A quick internet search suggests that annual production is currently about 200,000–300,000t. According to the Natural Cork Council,[4] about 25% by weight is for wine bottles, which accounts for about 70% of revenue.

If all current cork production was diverted to making cork housing, we could produce about 18,000 of these 44m² homes a year worldwide, however, we would have to find presumably less-sustainable replacements for the current use. The argument made by cork house proponents is that only waste cork is used. Even if there was such a thing, we would hope it was a small proportion of total cork production. In fact, the cork industry claims to be waste free – even cork dust is a valuable product.

So, plant more cork oaks! In 40 years, we can start harvesting the bark (or perhaps sooner with extra irrigation). Having the personal experience of planting hundreds of now dying ash trees, I will let others pick up the thought experiment and think of other issues that might make us rethink not just this project but our very approach to deciding what makes a building sustainable.

Our point is not to pick on one well-meaning project and rubbish it, but to use it as an example of a lack of critical thinking when it comes to what might be called an architectural response to, in this case, the climate emergency. If we don't debunk our mistakes, we risk all projects being tarred with the same brush by climate change cynics as eco-nonsense.

For the avoidance of doubt, we love cork. We hope that more cork oaks are planted and that the industry can thrive. It is an extraordinary material with the potential to be a genuinely sustainable product. We have used cork-based insulating plasters for renovation, rigid cork on flat roofs and even as friction material in motorcycle clutches. But as with timber, it is too valuable to waste. We must use it for what it does best.

ABOVE This cork-based insulating plaster is a very robust solution for internal (or external) wall insulation on old stone buildings like this one.

Design from nature

The Biomimicry Institute explains biomimicry as follows: 'Biomimicry is a practice that learns from and mimics the strategies found in nature to solve human design challenges – and find hope.'[5] This is intuitively appealing, and as such we should be wary.

If the natural world was the product of an intelligent designer, as some believe, then we should certainly find plenty of great design ideas to copy without having to reinvent everything. Take flight, for example; a problem solved even for larger vertebrates about 215 million years ago when pterosaurs took to the sky, some as large as small aeroplanes. Humans spent thousands of years jumping off cliffs and towers trying to fly like birds and it was only when we stopped flapping our arms that progress was made. From the first powered flight to the Spitfire took about 30 years, from the Spitfire to Concorde took another 30.

Having got off the ground, nature has provided design inspiration such as the swift-like folding wings of some fighter jets. But even the most successful of these is now retired from service. Copying nature with human-made hinges proved to be complex and expensive. Perhaps we will see bio-inspired mechanical muscles and joints in high-tech fighter jets of the future, but in the humble world of building there may be simpler solutions than trying the emulate billions of years of evolution.

It is likely that humans will struggle to replicate the efficiency and self-repairability of muscle, bone and joints, but it is even less likely that evolution will produce the self-evidently useful wheel or ball bearings. That termites can build mounds that can regulate temperature by controlling airflow is genuinely mind-blowing but these may not be the best inspiration for ventilating our buildings, any more than we would adopt the honey bee's waggle dance to give someone directions to the supermarket!

As with vernacular design, it is the process not the visual outcome that is of real interest in terms of learning about design. Evolution-inspired algorithms are driving AI and may have a role in design and other problem-solving, but we will avoid pontificating on that subject (not least because what we write today will look very out of date in a year – or even a week).

Design in nature is a very slow and blind process driven by random mutation and natural selection. Although we mostly see nature as beautiful and good, these are simply emergent properties of a value-free process that also produces flesh-eating bacteria and zombie parasites. In the garden of Eden not all flowers smell of roses; some smell of rotting flesh and are pollinated by flies.

So, while a well-woven story about inspiration from nature may well win you a first at university, we remain sceptical about biomimicry as a primary approach to sustainable design. Certainly do immerse yourself in nature and find time to walk in the mountains or listen to the waves, but the sustainable insight may be to simply build less rather than to 3D print a building as a naturally ventilated termite mound.

ABOVE Grain store in northern Spain built using natural materials readily to hand.

OPPOSITE The Elytra Filament Pavilion by the German 'experimental architect' Achim Menges, 2016. A robot mimics flying beetles as it weaves a canopy over the courtyard at the Victoria and Albert Museum, London.

Early responses to environmental concerns

A number of initial ideas about what 'eco' or 'green' building might be like have included several blind alleys that have proven to be ineffective, costly or difficult to reproduce at scale:

- Collecting rainwater to flush the toilet – requiring costly and unreliable pumps, filters and tanks, as well as energy to operate. It turns out that the supply of mains water is more economical, reliable and has a lower environmental impact. In remote rural areas, a borehole will usually be able to supply potable water.
- Composting toilets – a good solution for sites with limited water supply but too niche for most buildings.
- Reed bed treatment of sewage – requires space and maintenance. Only appropriate in rural areas.
- Passive stack ventilation – requires costly large cross-section, vertical stacks but ventilation rates are at the mercy of very variable wind conditions and temperature differences, which complex controls do little to avoid.
- Earth tubes to preheat ventilation air in winter and cool it in summer – large and costly underground ducts that require space and extensive groundworks, with additional hazard from rodents and mould.
- Individual wind turbines – need height and good wind speeds to be effective, are expensive, noisy, only generate small proportion of typical domestic electricity demand and are only really useful in remote rural locations where there is no mains electricity.
- Wood pellet boilers – expensive, unreliable and require large pellet storage with dust control to prevent spontaneous ignition. Arguably a renewable but very limited resource and definitely not low-carbon fuel. (This is discussed by Alan Clarke and Nick Grant in their paper for AECB, 'Biomass – A Burning Issue'.[6])

We have direct experience of many of these apparently sustainable measures – indeed, Nick is proud of his composting toilet – but we conclude that they have at best very limited application for enthusiasts only and in remote areas. Some, such as passive stack ventilation and micro wind, simply failed to perform as hoped, while others such as rainwater harvesting systems (as discussed in Judith Thornton's article in *Green Building Magazine*[7]) and biomass heating have been shown to be less sustainable than existing technologies.

Rather than further considering what can be viewed as dead ends in this book, we would prefer to suggest measures that our understanding and experience indicate are effective, present value for money and reproducible, at least at the time of writing.

OPPOSITE Strata SE1 stands as a monument to greenwashing. Building mounted wind turbines should not have made it past the most basic feasibility check.

ABOVE The original brief for the De Montfort University's Queens Building actually called for innovative solutions. The passive ventilation system has been troublesome and required very active control. The performance has been well documented by Bill Bordass and Adrian Leaman of the Usable Buildings Trust.

INDIAN Coffee HOUSE
ഇന്ത്യൻ കോഫി ഹൗസ്
Maveli Cafebru
RAMCO
SUPER
PLASTER
Higher
Coverage!
CPI
MAVELI CAFEE
INDIAN Coffee HOUSE
ഇൻഡ്യൻ കോഫി ഹൗസ്
POLICE
THIRUVANANTHAPURAM CITY

Problem-solving designs

The design process needs to address the characteristics that make buildings useful - that is, well grounded in their physical and social context; not self-referencing objects or icons but buildings that emphasise solutions to practical issues in preference to aesthetic judgements or the clichés of style and fashion. The designer must avoid placing self-expression above other considerations.

Throughout the design process, it is necessary to involve the building users and consider the social usefulness of the project in order to meet the needs of the building users, ensure good value and reduce costs without reducing performance. creating a place that is loved and cherished, for then it will be looked after, defended and improved over the years. Too many buildings appear to be designed to impress the designer's peer group rather than built around the users' needs and desires.

The first important step is to define the problem: to be clear about the objectives, map the constraints that give shape to it and think of possible solutions. It is important to view constraints as a positive driver of the design process, not as unreasonable or unnecessary limitations to thinking and creativity. They define avenues that are irrelevant or unhelpful to follow and they also suggest directions that will lead to an answer. Problem-solving starts with problem-finding – searching for things that are suboptimal and in need of improving; not just how but why should we solve this problem.

The budget is particularly important. It is crucial to clearly define the construction budget within the overall project budget and to identify the priorities. Too often, costs come into focus at the tender stage, which is too late to achieve an optimal outcome.

So, what features would a good design process include? Sustainability will be a given. This implies minimising energy in construction and use; designing for a long and useful life, well-built using robust materials in an adaptable manner; and minimising resources, waste and pollution. There is often an assumption that sustainability is some sort of aspirational add-on. This cannot be the case. Too often, we hear designers say that they would love to do a sustainable building but the client is not interested. This is like saying we would like to make a structurally sound building but the client has no interest. Once embraced as constraints, we will see genuinely sustainable design.

Balancing aspects of the design

A good design has various aspects - orientation, form, plan, performance and quality, to name a few - in balance with one another. No one aspect has unreasonable importance in the overall scheme of things. The problem is less that one aspect has been done too well; more that it is likely that other areas have had insufficient attention.

If there is a conflict between the aesthetics and function, then this is most likely to be due to fixed ideas about what something should look like rather than a false dichotomy between a building working well and looking good. It makes little sense for the design to be hostage to subjective and changing ideas about how things should look. A visual image is not a good indicator of practicality or value. Time is a good test of a building and can give the impression that old designs are often better when, in fact, the opposite should be true as knowledge and experience build over time. What we see is the once-new objects that have stood the test of time becoming classics. Sometimes designs are unfashionable when new but earn classic status with the eventual recognition of a great design that was ahead of its time.

An inclusive design team

This implies that the various specialisms in construction, structural and services engineering, for example, are fully integrated into a design team. Our experience includes too many cases of mechanical services being 'retrofitted' into a design concept that has not taken the constraints into account; for example, space for a plant room is often insufficient, which leads to poor performance and high cost.

For this 'inclusive' approach to work well, it is necessary that the members of the team are all familiar with the considerations pertinent to the various specialisms of others as well as their own particular area of expertise. There is significant advantage to be gained from established clients, contractors and design teams being able to work together on successive projects, carrying forward the experience and learning from one project to the next. The benefit of this approach cannot be overstated but most tendering rules for larger projects make it very difficult. Competitive tendering is intended to provide value for the client, though it is not always the best way of achieving it.

The above implies in turn that design education should have far more emphasis placed on interdisciplinary working. This is in contrast to some current architectural courses that have little in-depth teaching of structures, environmental science and building construction.

A circular design process

The design process is not a linear, proceed stage by stage to completion, kind of activity. One might expect to start with an initial idea that you then elaborate, develop and refine until you reach a resolved final design. We find, however, that the process is more like a circular journey spiralling towards resolution by way of a number of repetitions that cover the same ground but with additional information and clarity added with each repetition.

BELOW HEM Architects in Sheffield take collaboration seriously. Each consultant and the client is a valuable part of the team and both models and drawings are used to discuss ideas effectively.

In this way, an initial sketch idea would be formed by considering the aims and constraints at a very outline level – orientation relating to context and climate, form and layout, and the fabric of the building, including strategies for limiting the usage of energy, structure, construction and services. What is important at this early stage is that initial ideas are not limited to what things look like, an external 'massing diagram'; do not be tempted to shortcut the design process by jumping to a premature visual conclusion, but consider all the important features of the proposal including non-visual aspects at a preliminary level. If designers do not feel compelled to reinvent themselves for every project then that iteration can evolve over many buildings, with valuable feedback from the build process and final use. If they find something that works well, resist the pressure to change it on the next project just for the sake of change.

Adding detail

These aspects of the design would then be developed by analysing costs in detail, feeding in more information on needs, regulatory requirements and so on, leading to a preliminary proposal with a plan. This would then have detail added and additional considerations taken into account – initial analysis of the structure, for example. This is then repeated with more detail and information fed in to arrive at a proposal that shows the arrangement and size of all the main elements.

Another iteration would take that proposal and add further information and detailed consideration of how it is to be built. This would include the availability and cost of materials, which is then fed in to create a complete design.

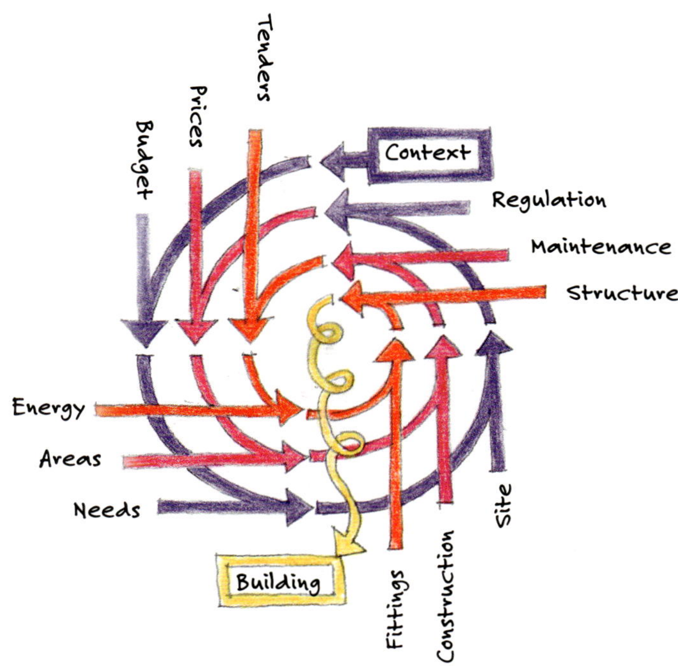

LEFT Design is not a linear process.

The role of the contractor

The role of the contractor is critical at this stage, though competitive tendering can make it difficult to involve a contractor during the development of the design. Contractors are also often reluctant to invest time in what they may see as a process that may not come to fruition as a real construction project. Instead, one has to fall back on experience and knowledge of practical building; however, not all designers have this understanding in large measure (see Chapter 4.5 for ways in which to involve the contractor at the design stage).

What will it be like?

It is important to establish an idea in one's head of what a place will be like; to take an imaginary walk around the building and envisage how it relates to other buildings and the landscape around it. Consider what it is like inside as well as out; have an imaginary look inside, look out the windows and consider what you will be seeing. Then take another virtual trip outside - look back at the building and consider where the windows will be and how the external appearance will be organised. Not all clients or designers have this ability; even if you do, it is useful to seek out similar spaces to visit. This could be a single element such as a window or an entire room with a similar function and location, allowing a direct comparison.

As the design develops and more detail is added, imagine how you would actually build it:

- In what order do the pieces fit together?
- How would you position them in space on a building site with few fixed points to refer to?
- How are they fixed?
- How are they are adjusted to ensure a good fit visually as well as practically?

RIGHT A Segal house under construction. At this stage one can look out and see the view, the direction of the sun, and decide on the position of the doors, windows and non-loadbearing walls. It is a special case as generally buildings are built from the bottom up unlike a Segal building where the roof goes on as soon as the frame is up.

This understanding of construction helps to devise buildings that are straightforward to put together, which leads to speed of construction, good value for money and a high-quality finish.

Fundamental principles

It is a good idea to consider the basic elements of design that can achieve radical simplification and save time, money and resources. For example, overlapping the use of space with circulation through rooms instead of a dedicated corridor enables a larger room.

Similarly, at a more detailed stage when considering the construction of the building, simplification can follow from thinking through the fundamental requirements of the elements of construction. Architect Walter Segal was a master of this approach – simple, but not too simple. He devised foundations for self-build applications that did not require the site to be levelled; flat roofs that are reliable and simple to construct and stairs that could be made without cutting any angled timber (the whole construction only requiring simple right-angle cuts). Requirements for fire and structural safety as well as thermal performance have changed considerably over the years and so many of Segal's ideas are now too simple, so a redesign from basic principles fully embracing the new constraints is required.

LEFT Simple Segal foundations – a concrete-filled hole in the ground.

RIGHT Jon took the hard work out using a modern post hole auger.

The search for quality

Attitude to work is significant in this context; are you intent on doing the best you can, doing a good job for the sake of it or aiming to do the least amount possible? Are you an idealist or a pragmatist? Do you strive for perfection or have a more relaxed approach to life? Do you seek the emotional rewards that skill and competence can provide or are you looking for a sense of worth in other ways?

We believe that producing good quality needs care – people who care about what they are doing. This is easiest to achieve in a collaborative working environment where people talk to one another and work things out together, whether in the studio or on the building site, as this attribute applies both to skilled manual work and abstract thinking. There is a direct link between hand and head; you need to think through what you are doing and how to do it in the most effective way.

A skilled worker thinks the whole process through from start to finish – what materials are required, what tools to use and so on – whether it be drawing a plan or digging a hole. It is important to recognise that manual work requires skill and is not mindless labour. Furthermore, a skilled builder will be able to perform tasks quickly – they are anything but slow and expensive, in contradiction to the assumption that craft and quality require additional time and money.

Motivation, aspiration and the positive feeling that comes from competence are key factors in creating quality, creativity and talent, less so. We all possess the basics to do a good job. Experience has shown that almost anyone can build a house if they put their mind to it, even if you need help from friends or family to do the more labour-intensive tasks!

Many builders have a desire for quality, which can be brought to bear on small projects. For large projects, it requires establishing a trusting collaborative relationship between designer and constructor forged at pre-contract design and post-contract site meetings. The tradespeople actually doing the work are generally very good at solving problems on site but subcontractors working within tight budgets and short timescales are not necessarily predisposed to buy into a discussion on quality. Toolbox talks are one way of establishing an understanding of the organisation from top to bottom, but success relies on the attitude of the people involved more than anything else.

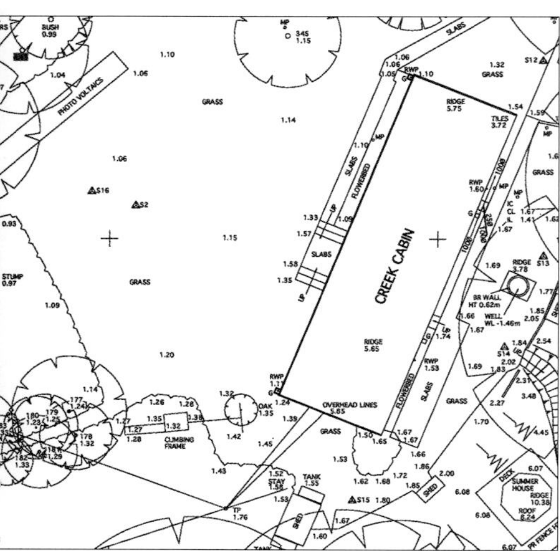

ABOVE Accurate digital drawing
for constructing a building.

The role of drawing

Architects, designers and engineers produce information in the forms of drawings and specifications for contractors and builders to interpret. So, how should that information be presented to ensure that their intentions are clear? Designers need to be digitally fluent in producing 2D and 3D technical drawings and presentation images. However, they also need to be skilled at sketching using a pencil and paper as a way of thinking creatively, exploring ideas and developing alternative solutions to problems and constraints. Hand drawing on paper rather than using a mouse onscreen will allow you to work your way into the problem and its potential solutions. Allow a measure of incompleteness and lack of resolution. This prevents opportunities from being closed down too early.

Generally speaking, young designers these days tend to go straight to the computer rather than the sketchbook when they start to design, which often shows in the buildings that result. Less time is spent developing ideas and more is spent visually enhancing first thoughts. Computers can turn complex curved forms into mathematically reproducible code, which can then be given physical form on the building site, enabling the construction of dramatic but expensive buildings. They can also produce accurate views of places and buildings from any angle. Do not overuse bird's eye or other unnatural viewpoints, but do make sure there are street views taken at eye level.

Limitations of the perfect solution

Avoid pursuing a problem to the point of obsession, which can result in tunnel vision to such an extent that you exclude the very ideas that may lead to a solution. Experience comes from making mistakes over and over again but with more confidence. Seeking the perfect solution can lead the designer to self-consciously show off what they can do rather than developing what the design sets out to achieve. Drawing attention to how important, unusual and innovative the design is often gets in the way of taking advantage of the opportunities that present themselves to develop a practical solution. The trick is knowing when to stop.

Professionalism

How we approach work and our attitude to quality reflects on the role of professionals. In the early nineteenth century, the idea of professionals as impartial experts arose as intermediaries between rapacious magnates and powerless consumers. The professional doctor or lawyer built trust based on practical advice and promoted fair play to established standards of quality and business practice. As the century wore on, this role became skewed as professionals became allied to the ruling elite in the exploitation of less educated and less wealthy citizens.

The twentieth century saw professionals slowly becoming concerned with improving standards of quality and service to citizens. This led in the latter half of the century to many professionals being employed by the state in the civil service for the public good: doing research, providing advice, and establishing best practice and minimum standards. In the building industry, for example, the Building Research Establishment, the Fire Research Laboratory and design offices in the Ministries of Health and Education were all set up to develop standards for good, safe homes, schools and hospitals.

A lack of care

Some of this remains, although outsourcing government activity to the commercial sector means that the government relies increasingly on private companies for advice and other services. Professionals are now often employed by these companies to sell services and products.

These companies are able to extract considerable fees from public funds and sometimes have loyalties not wholly dedicated to the public good. The corruption exposed within the suppliers of cladding to the refurbishment of Grenfell Tower and the lack of regulatory oversight are exceptional but indicative of a general erosion of standards and responsibility within the industry and a lack of care at many levels.

ABOVE Quality is hard to define but we know it when we see it or when it is absent.

LEFT Free-form sketch for developing ideas.

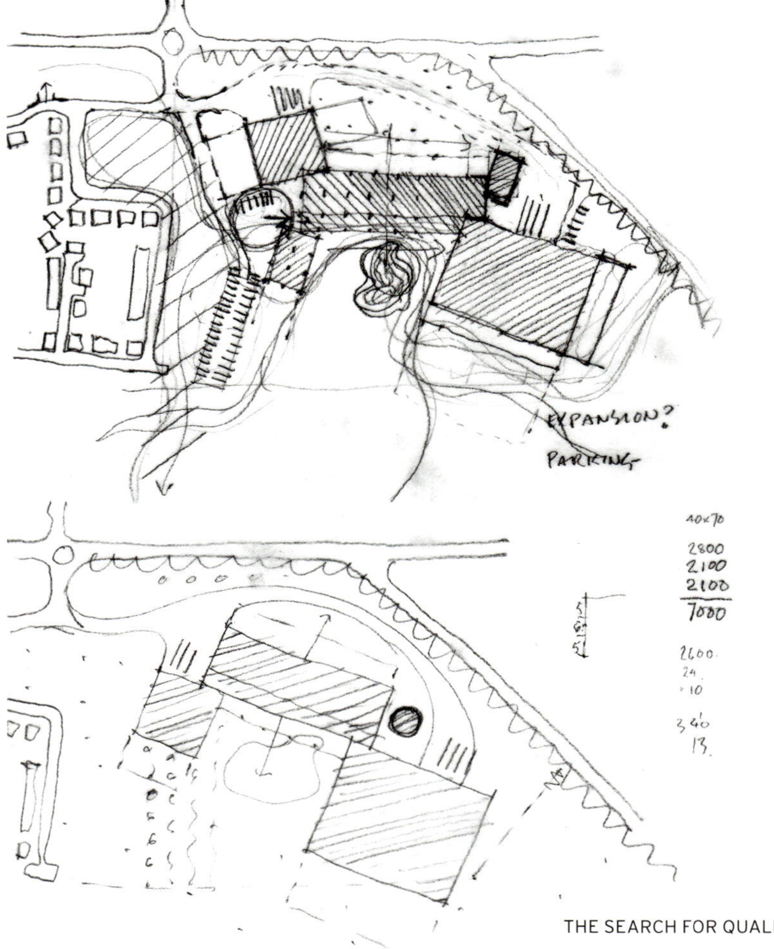

1.7

A modern vernacular

LEFT Romily, self-build by Brian and Maureen Richardson.

Consider vernacular building; that is, the vast majority carried out without professional involvement, using a wide range of readily available materials and unselfconsciously reflecting local traditions and cultures. Our thoughts are the product of trying to make sense of confusing but widely held attitudes to buildings and architecture. A nostalgic vision of thatched cottages has formed the basis of a 'Tudorbethan' style, where many people believe that modern architecture is bad and tradition good, and the public view is conditioned by advertising, TV, social media, fashion and what is available at the DIY superstore.

How can we create good buildings within this context? Buildings that are devised with quality of idea and thought, substance and execution. Buildings that are comfortable to use and inhabit, fitted to their environmental and visual context, technologically adapted to the requirements of use and economical to build. Buildings, in other words, that emulate the best of traditional vernacular construction.

What is important is the *process* by which vernacular buildings come to be – not so much how they look or what style they follow.

Vernacular building in the past

In the past, vernacular building was the product of commonly understood elements, methods and materials passed down from generation to generation, person to person. A simple and accessible technology provided a language of building, within which each building tells a different story. The subtle and infinite variety of buildings evident in our old towns and villages is an expression of social needs and site conditions within the common language. The building process was in the hands of individuals able to exercise choice to suit the building to their needs and wishes within the framework of established practice. Techniques of building varied from one area to another and were based on locally available materials, used to a large extent in their natural state with a minimum amount of fabrication. Transporting materials was difficult, so a technology developed in each area based on the particular materials available.

Vernacular building was not the result of a set of stylistic or ideological values, but rather the product of the practical process of people building for their own use in an economical manner. It is the product of the builder's direct first-hand response to existing buildings, landscapes and the microclimate – the experience of places that not only function well but are also full of life from the human activity that takes place there, the rich source of potential experiences of all kinds. Each building and detail demonstrates the skills and personality of those involved. This is the antithesis of the sterility of so much modern design and urbanism.

The products of this process display many features we regard as valuable, including human scale, richness of visual texture, economy of means, variety within order and an inherently environmentally sound use of materials.

The vernacular now

This pattern of development was largely swept away by the industrial revolution, which placed development in the hands of industrialists and financiers building for profit rather than commodity, and using mass-produced, manufactured products made from materials transported from afar. Individuals were no longer part of the process. They had been distanced by the increasing complexity and specialisation of building. As in other fields, this has rendered building the province of the professional designer working in their office.

Most housing in the past has been designed and built without architects being involved. However, reconstruction and development since the Second World War has often not created good places to live, work or play – streets, neighbourhoods and towns that are affordable, durable and democratic. Planners do not have strategic aims setting out land use, scale and other constraints. Streets that were once safe spaces to meet, play or host street parties are now dominated by noisy, polluting vehicles.

This book is not concerned with looking back to some fictional golden age of traditional material and technique. Neither do we believe in the archetypes of a traditional or classical response. Stylistic concerns of this kind spawned the vernacular revival of the early part of the century, which had expression in country houses for wealthy middle-class clients and more recently Poundbury, a new village addition to Dorchester sponsored by the then Prince Charles as a critique of modernism and designed by Leon Krier, an urbanist trained in Germany. The fire station is a pedimented neo-Georgian palace with three garage doors stuck on the side. The whole place is more like toytown than a living community. Residents, who make places come to life, are no longer engaged in how places are created. We need to learn from the past, but not this parody of a modern vernacular.

LEFT Designed without architects and still a living and liveable place in South East London.

RIGHT Recent development close by with no sense of place and dominated by the car.

Four aspects of vernacular building

We believe that we should look to the process by which places are created. Above all, they should be useful, and to be useful to those who use them they must reflect their needs and wishes. To do that, users have to be involved in the process of their creation. Our interest is, therefore, to do with the manner in which people can become involved in planning and building. The aim is to produce modern buildings that are clearly of their time but draw on the traditions of the past; to apply the lessons of the past to current needs and methods.

The following four ideas come together to understand a modern vision of vernacular building:

1. Dweller control
2. Pattern language
3. The 'Segal method'
4. A sustainable future

1. Dweller control

The first idea of 'dweller control' was developed in 1976 at a theoretical level by architect John F. C. Turner in *Housing by People;*[1] the implications of which for UK housing policy have been developed in 1990 by Colin Ward, an environmental educationalist, in *Talking Houses.*[2] The theory suggests that mass housing initiated by a developer or politician, financed by bankers, designed by professionals to the requirements of the housing official or sales agent and built by a contractor, will never satisfy people's real needs and will always be inherently uneconomical for residents.

Importantly, this idea presents dwelling as a process, an activity, and then argues that the user should be involved in that process to ensure usefulness, economy and sustainability. This may seem like a utopian idea, perhaps limited to a few self-build schemes. However, even buildings that are being designed before the occupants are known can take into account the varying needs and desires of future occupants. For example, Spreefeld is a co-operative development in Berlin that has open-plan floors that can be subdivided to suit the needs of different households. We consider this in more detail in Chapter 1.8. Flexibility and adaptability are key.

2. Pattern language

The second idea is Christopher Alexander's notion of a 'language' of 'patterns',[3] which describe the elements of the built environment. These are widely shared and understood and form a vocabulary of common techniques and details which, using local materials, produce traditional vernacular buildings.

Patterns are the building blocks that go to make the rooms, buildings, places, towns and cities around us. They are not concrete elements like bricks or doors but rather rules of thumb that suggest elements that can create good places and buildings rather than bad ones. These patterns attempt to make explicit those invariant features and attributes of places and buildings that work well and which we like, often without being able to articulate why. They can be used to create an infinite range of examples, all different but all sharing those essential features that make them function well. A pattern language gives each person who uses it the power to create an infinite variety of new and unique buildings, just as everyday language gives a person the power to create an infinite variety of sentences.

Alexander and his co-authors sought to make the assumptions of good practice explicit and to organise these ideas according to scale. *A Pattern Language* has clarified the values implicit in our thinking and can provide a useful framework for analysing ideas with which everyone is familiar but most people no longer feel able to manipulate. Decades after its publication, it is still one of the best-selling books on architecture.

Why are some buildings and places loved above others?

An essential ingredient of good buildings is for them to remain wanted, so they will be protected, cared for, altered and improved over the years. But what is it that will make them appreciated? Is there a special quality that makes a good building which means some buildings survive, where others are disposed of before their time? We believe it to be a combination of beauty and adapting to changes of purpose.

Patterns have been lost

These patterns are repeated everywhere and make up the world we know. People have developed different languages of patterns in different places, which reflect various circumstances and cultures. However, one of the difficulties we face is that these pattern languages have broken down and been replaced by a postmodern attitude where anything goes. People no longer share a common understanding of the environment we inhabit.

Consider the pattern Christopher Alexander calls the 'window place'. There is something objectively wonderful about a room with a bay window or window seat in it. The feeling that rooms with these kinds of features are especially appealing is not whimsical. It is based on our attraction to light and sun, and the need to be able to sit down and make ourselves comfortable.

However, a great many rooms we inhabit have no 'window place' – rooms where the windows are merely holes in the wall with no special place to sit, bathed in light, and watch the sunset. This creates a tension between being drawn towards the window but at the same time searching around for a 'place' to settle in the room. It is uncomfortable. It becomes clear why some places are so much more pleasant than others when we start to analyse these patterns.

Finding our own pattern language

The results from following individual pattern languages will vary enormously and will mirror the particular priorities and features that appeal to individual taste and circumstances, the particular physical and economic context and the materials to hand. Some classic, some quirky, but all addressing utility and delight.

Working in Vancouver, building design consultant Monte Paulsen's work identified a

ABOVE Pattern 180 from the Pattern Language outlines the need for a 'place' to sit which could be a bay window, a window with a low sill, an alcove by a window or a built-in window seat.

A PATTERN LANGUAGE *from* PASSIVE HOUSE

New patterns from old ideas

No 1: LAYOUT MECHANICAL FIRST *
No 2: MINIMIZE SURFACE AREA **
No 3: COURTYARDS THAT COOL
No 4: FAT STAIRS, SLIM HALLS
No 5: TRAIGE BALCONIES *
No 6: ARTICULATE THE SPACE BETWEEN
No 7: REPEAT WHAT WORKS **

RIGHT (BOTH) Examples of Monte Paulsen's work at a presentation.

number of patterns that can help in designing cost-effective, energy-efficient buildings. As of writing, Paulsen's work is unpublished and sadly he has since passed away, though his research can be accessed online.[4] Many of Monte's patterns contradicted those of Alexander as they reflect modern concerns and technologies. For example, Alexander delights in small window panes which make no sense with modern double- and triple-glazed sealed units.

Our own self-built homes and professional design work owe a lot to the pattern language books and approach; however, the pattern language has limitations that you will have to be aware of and modify. The idea implies a certain folksy aesthetic through the illustrations and it becomes less and less useful as it moves from considering large-scale towns and neighbourhoods through planning individual buildings to descriptions of how to build, which are generally too crude to satisfy current standards of construction. However, we treat it as a useful tool based on empirical and subjective observation and not as a set of physical laws nor a belief system. In its published form, it makes many assumptions about people's outlook and ways of living, and it disregards some social, political and economic realities. Some of the patterns are mutually exclusive, while others are clearly dated, as we might expect from books written in the 1970s. Nevertheless, it can be a useful way of getting people involved in these issues and able to develop a shared vocabulary that could lead to a modern vernacular approach.

3. The Segal Method

Just as pattern language helped with issues of design, so the 'Segal Method' described by Alice Grahame and John McKean[5] of timber construction reconceived the process of building from first principals. This idea allowed ordinary people with a basic understanding of radically simplified construction to become active participants in the step-by-step process of developing a design. The dimensional discipline of a planning grid is easy to manipulate and the post-and-beam frame allows openings for doors and windows to be positioned at will. A Segal Method building is very adaptable and therefore likely to have a long, useful life, just like a traditional vernacular building.

While the rational, economical Segal Method opened up opportunities for people without previous experience to build, not everyone wants to build their own home; most people move into existing dwellings. Although the Segal Method has proved to be difficult to update to meet current standards of energy performance, the radically simplified 'dry' construction based on the use of readily available materials used in their standard sizes within a modular grid which defines the planning, structure and construction, has useful lessons for current practice.

4. A sustainable future

This leads to the fourth and final idea: we must work towards a sustainable future. This requires:

- energy consumption for heating and lighting to be drastically reduced;
- buildings to be adaptable to changing requirements;
- toxic materials and polluting processes to be phased out;
- non-renewable materials to be eliminated.

In addition, as buildings become more energy-efficient in use, so the energy required for their construction becomes more and more significant.

Traditional vernacular buildings were inherently 'green' because materials were generally used in their natural state, with little or no processing, and of local origin. In addition, heat was provided by burning wood, a renewable resource at the time. A modern vernacular, similarly, should utilise low-carbon products as far as possible. However, it is difficult to locally obtain the materials necessary to build. Another feature of an ecological vernacular is that the design of buildings would be responsive to the particular context in terms of orientation, shelter and landscape, just as these were the concerns of vernacular builders in the past.

We believe in the value of variety but within a framework that provides an overarching order. In this way, the use of a repeated simple building type does not create monotony. Irregularity of the terrain or deviation from standard measurements results in small variations which strike a balance between unity and diversity. This variety can apply to the basic ingredients of a building: material, texture, colour, proportion and so on.

ABOVE Pisticci, Italy, variety within an overarching order.

Modern vernacular building

We believe that dwellers in control of their destiny - with a shared way of thinking about building and accessible methods of construction - working towards a sustainable future, can create a modern vernacular architecture.

This shares many attributes with traditional vernacular building:

- human scale;
- community involvement;
- design with climate;
- relationship with place and landscape;
- the use of readily available materials;
- adaptability;
- individual designs, which are not mass produced;
- diversity within an overall order;
- accessible building techniques;
- creativity arising from constraints and inventive by necessity;
- economical but high value;
- character and personality.

Such a modern vernacular does not necessarily use local or traditional materials. Neither are the forms and details directly related to traditional forms and methods of construction. For these reasons, many critics would not describe such buildings as in any way 'vernacular'. We believe that this is to miss the point because what is significant is the process and the involvement of people in the process (for more on this see the writings by Colin Davies[6] and Graham McKay[7]).

Integrating those characteristics of earlier forms of development that work leads to evolutionary development rather than merely change for change's sake. The challenge is to develop an approach that creates a culture of building that is commonly understood and forms a role for people in the creation of buildings, places and landscapes.

OPPOSITE Black house by Meredith Bowles, Mole Architects. With modern buildings the wall cladding is rarely structural so stone and brick make less sense as a simple rainscreen. Some architects 'reference' the vernacular whilst a growing number are learning from the approach.

The modern vernacular in action

The previous chapter introduced some theoretical ideas that are essential to provide a guide for action. Next we are going to discuss how one can translate that theory into practical solutions and buildings, without which any theory is of very little practical value. These ideas have been around for a long time; there was a 'movement' in some architectural circles in the 1970s for 'community architecture', and it feels as if this way of thinking has particular relevance to current housing policy.

Before getting started, it is important to consider the concept of sustainable development. The Bruntland Report[1] defines this as ensuring that the needs of the present are met without compromising the ability of future generations to meet their own needs. Jon prefers a more specific definition – to pass on to future generations a stock of capital assets that will allow the quality of life to be maintained in the future.

One can then look at what those assets might consist of:

- *Social* assets – the idea that we're in the business of social development and inclusion.
- *Human-made* assets – buildings that need to be adaptable so they meet the changing needs and expectations of the future.
- *Natural* assets – natural resources such as fossil fuels, metals and water.

In the construction world, this suggests that people need to be involved in the major decisions that affect them, that buildings need to be adaptable to changing circumstances and that natural resources of materials and energy from fossil fuels need to be conserved.

The Whole House Book: Ecological Building, Design and Materials by Cindy Harris and Pat Borer of the Centre for Alternative Technology has, as its title suggests, encyclopaedic coverage of the design of sustainable homes.[2] Here, we concentrate on the role of building users and adaptability.

Participation for sustainability

We argue that people and their participation in the housing process is a necessary pre-condition for a sustainable housing process. This is in clear contrast to the generality of housing in Britain, which relies overwhelmingly on mass housing provision, the production of which has no involvement with the people who live in it. This goes back a long way, from the creation of the Victorian industrial city; the involvement of the philanthropists and later the local authorities in mitigating the worst conditions created in those industrial cities; speculative developments between the wars; and more recently, local government and housing association involvement in providing housing. On the whole, none of this history involves residents.

Participation in housing has particular resonance because we have a very intimate relationship with where we live, where we call home. Jon's background is not exclusively in housing, but a large part of his work has been in that field with particular involvement with people designing and building their own homes.

This practice is at the top of the ladder of participation suggested in the 1960s by American urban planner Sherry Arnstein.[3] It starts on the bottom rung with manipulation – some would say deception even – and rises through information giving, consultation perhaps, then partnership. This is a fashionable idea these days, but rising to the dizzy heights where people can have power to control the process that they are involved in.

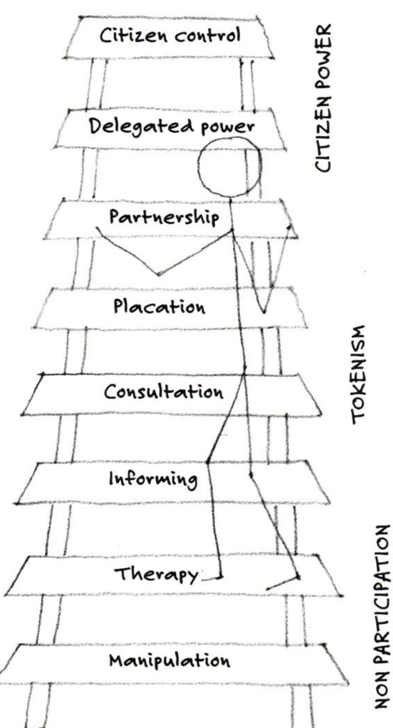

ABOVE From manipulation to control.

RIGHT Self-built back yards in the centre of Delft - variety within a planned framework with some of the visual texture of a long-established neighbourhood.

Dweller control

The idea of dweller control was developed by John F C Turner whose first Law from *Freedom to Build*,[4] published in 1972, suggests that when residents have control of the major decisions affecting the design, cost and management of their housing, then both the process and the outcome will tend to satisfy their needs and values unlike the usual housing situation where residents have little or no say if the resultant homes and environment act as an impediment to wellbeing.

And thus, a burden on the economy and unsustainable in the longer term. It's an underlying idea that seems to us to be fundamental to what we're talking about.

Turner went on to suggest two other propositions, which are also relevant in this context. The first is that the most important thing about housing is not what it is as a physical object but what it does to people emotionally, the value they place on it, how much it costs for them to live in it and so on. This distinction is sometimes difficult for designers to appreciate. We spend time worrying about what things look like and not nearly enough about what they do, how useful they are and what value they have.

The second preposition is that deficiencies and imperfections in one's own housing are infinitely more tolerable if they are your own responsibility – if you were the person, in other words, who is accountable for the decisions that created a deficient commodity – and, let's face it, nothing is ever perfect. Building is about a balance between dreams and reality, between brief and budget and what can be achieved in the time and within the regulations, and so on. And so, a participatory process – a process in which people are responsible for the decisions along the way – is likely to create buildings that people have much higher levels of satisfaction with. This is a fundamental idea, and one which by and large is not recognised in the way housing is developed.

The late anarchist writer, Colin Ward, wrote extensively on environmental education and sustainable housing. In his 1985 book, *When We Build Again*,[5] he pleaded for 'housing that works'; housing policy that is not paternalistic or authoritarian but which recognises that people are not helpless and inert consumers but rather in possession of the ability and desire to shape their living environments.

His proposition is that the opportunity to participate in this process is a necessary part of a democratic way of life.

Accessible construction

As we outlined in the previous chapter, the pattern language has limitations, but nevertheless, we have found the idea a very useful way of thinking about what makes good buildings and places. It enables us to consider our own language of patterns, our preferences, and enables us to think about how to turn such preferences into real designs and buildings. It is a great way of considering design with people who are unused to thinking about what makes a place good.

Having a framework within which we can think about how houses and neighbourhoods are planned, the next part of the story is having a way of building that is accessible to people who are unskilled in building construction. This is where modern self-build pioneer Walter Segal comes in, with houses built using standard components bolted together within a post-and-beam frame, and organised within a dimension framework developed with the standard sizes in which those components are available. This has a number of significant attributes, one of which is that the walls are non-load bearing, so posts hold the building up instead of walls. In this way, you can move walls around to where you want them and decide on where you want windows and doors on any floor of the building. People who are unfamiliar with reading plans have found this handy because they can stand in the partly constructed building and decide where the view is, where the sun comes from, and where and how large the openings should be. This is in distinct contrast to the conventional way of building, in which every last detail must be established before work can commence.

Ken Atkins, who was the chair of the first Lewisham self-build housing association, put it like this:[6]

We must be the first council tenants who have been involved with an architect in the design of our own homes. The architect used graph paper to help us to represent the modular concept of 2 feet 2 inches, and asked us to draw a house within the cash limits. This was about 100 sq. metres. We did this as a group, and then took this to Walter Segal's house. He took all the ideas, and drew up around 50 or 60 house plans and then we went back as individual families to choose and adapt our designs. Every wall is non-load bearing so it is adaptable and changeable at any time during the process of design or building, or after I have lived in it. If I feel I want to change it, I can take out any wall and change it.

There are not many people who feel that confident about rearranging their house. From Walter Segal's perspective, he said at the time:[7]

Help was to be provided mutually and voluntarily, there were no particular constraints on that, which did mean that the good will of people would find its way through. The less you tried to control them the more you freed the element of good will. Children were, of course, expected and allowed to come and play on the site, and the older ones also helped if they wished. That way one avoided all forms of friction – each family were to build at their own speed and at their own capacity. We had quite a number of young

Well, it may be astonishing but it's there for sure. It wasn't a question of abdicating the control of the process by the designer, but rather was a collaborative affair with Segal making very sure that the structural integrity of the building wasn't compromised at any time. In Jon's experience, this makes it a lot easier than designing to an abstract brief for a

BELOW Debbie built her house from the ground up (with help from a friend) at Walter's Way, Lewisham.

ABOVE Self-builders get together
to raise the frames for Segal house.

client you don't know and will never meet. If you have people telling you what they want, it's pretty straightforward. We think this is something that some designers find difficult to understand.

Support structures

Here's another name from the past. Walter's beam-and-post frame has connections with the idea of support structures proposed by Nicholas John Habraken, a Dutch architect, in the 1960s and translated into English 10 years later as *Supports, an alternative to mass housing*.[8] He proposed 'support' structures – multi-storey frameworks within which individual dwellings could be constructed, altered and taken down independently from one another leading to a great variety of dwellings within the overall discipline of the support structure.

His principal aim was to propose an alternative to mass housing, creating a fixed framework of communally owned, permanent, mass-produced support, separated from the individual house under the control of the occupant. This idea of separating structure and other 'layers' within the building is quite a fundamental one.

This idea has more recently been elaborated by American project developer and writer Stewart Brand, who wrote *How Buildings Learn*.[9] In his research, he examined why some buildings improve with age while others are neglected until they are either demolished or fall down. It's an interesting question. He elaborated the original concept of Frank Duffy (an English architect working in the commercial sector), where the separation between layers in a commercial office building is often quite clear. What Stewart Brand did was identify what he called the Six S's:

- *'Site'* – the eternal, the fixed;
- *'Structure'* – there for at least 60 years, maybe 300 years;
- *'Skin'* – the envelope of the building;
- *'Services'* – note here the need to separate services from other elements of the building rather than to integrate them. This allows them to be changed within the lifespan of the building when a heating or electrical system becomes obsolete, within around 20–30 years;
- *'Space-plan'* – partitions;
- *'Stuff'* – furniture, the carpets, the curtains, *etc.*

This useful idea has been built on by Stephen Kendell and Jonathan Teicher in their book, *Residential Open Building*,[10] where they apply this layered approach to examples from across the globe. They identify in particular how the Japanese building industry has developed this idea, with factories now producing sub-assemblies for fitting out prefabricated steel and concrete shell structures. The Japanese have also developed flooring, partitioning, electrical, plumbing, bathroom and kitchen systems using tried-and-tested technologies. One can design and order a new house while sitting with a designer in a supermarket. Digital control of the production process allows prefabricated dwellings to be produced in infinite variety to suit different tastes, requirements and circumstances.

It was Japan that Sir John Egan turned to when researching *Rethinking Construction*,[11] which was the government's response in 1998 to the lamentable performance of the British construction industry. Egan's report sought to improve quality and reduce costs of building in Britain, and it identified prefabrication as one way of doing this. What you won't find in Egan's report is any mention of sustainability or participation. However, Michael Ball, who wrote *Housing and the Construction Industry*[12] around the same time, was critical of mass house builders. He highlighted the inflexible planning and construction of British homes as a problem: 'The mass housing industry in Britain, unlike that of Japan and elsewhere, offers poor value for money, over standardized products, poor standards of space and specification and an absence of adaptability.'

Adaptability and sustainability

Adaptability is important and tends to go along with the idea of people being involved, which reflects the social aspect of sustainability. Residents need to understand how buildings work and operate, and should be informed of how to ensure they keep operating efficiently – it's all very well reducing water consumption in a building by installing water-efficient fittings, but if you leave dripping taps running and don't maintain buildings properly, it's all to no avail. The point is that people have an essential part to play; it is said the whole issue of sustainability is 80% people, 20% technology.

It is also true that the people that we deal with when designing houses are generally very keen on saving the planet and saving money. In many cases we've found they're even more passionate about it than the professionals in the housing business, including the designers and managers. This was recognised at the Rio Earth Summit in 1992, which established Local Agenda 21: the idea that local people are an essential part of developing sustainable policies and making developments sustainable.

A sustainable housing policy

There is a danger that we will not create housing stock that is going to serve us well in the future. For this reason, there is a need to reintegrate some of the ideas from the 1960s, 70s and 80s into the current rhetoric of modernisation and efficiency in the house-building industry in Britain. The risk is that if you don't reintegrate adaptability and a participatory process in the drive for reduced costs and better quality, we're going to end up with

RIGHT River Studio, an architect's office reusing a very adaptable steel barn structure, Sjölander da Cruz Architects. Designed to the EnerPHit standard for energy-efficient retrofit of existing buildings, inspired by Passive House principles.

technological solutions (volumetric housing, to name one current idea) in the hands of the wrong people – contractors, financiers and politicians pursuing the wrong ends. This is why tower blocks, prefabricated concrete construction and other technological 'answers' have failed in the past.

There is an alternative vision around a modern vernacular; people who share a way of thinking about building and accessible methods of construction reinterpreting modern materials rather than using traditional materials – using steel for rainwater goods or a multistorey concrete frame allowing complete flexibility of layout, for example. We think this shares a number of features with traditional vernacular building in terms of diversity within some overarching discipline, low energy use and with a long, useful life ahead of it – a sustainable model for housing, in other words. What housing does for occupants is what really matters, and experience suggests that people have grown in self-confidence when working with the authorities, professionals and each other to design and build for themselves. This self-confidence fits very well with the current rhetoric of social inclusion and regeneration of urban areas, but effective participation in the housing industry remains very much on the margins and not – as it should be – at the heart of a sustainable housing policy.

These ideas should also be applied more widely in sectors other than housing. There have been some very successful examples. Peter Hubner involved teachers and children in the design of their new school in Moers, Germany, in 2014. In 1985, Cullinan Studio in London carried out a design process with staff and patients for the Lambeth Community Care Centre. Note that in both examples, *all* building users, pupils and patients included were involved – not just the professional teachers and doctors, as is generally the case in developing a brief for a project.

Self-build

Self-build is a term that conjures up different images in people's minds; from commissioning that dream house – a big detached property deep in the countryside with a large garden and swimming pool, to a do-it-yourself extension. To some it means buying a house built by a developer, but with various custom choices of layout and specification, or a kit house from a prefabricated house manufacturer. Some may join a group of like-minded people to build a number of dwellings together, possibly as part of a community-based project or private development. Alternatively, you can find a site and do it alone, either employing individual tradespeople or subcontractors or carrying out all or most of the work yourself.

What these approaches have in common is that they are an alternative to mass housing provision; they are usually very energy-efficient, individualistic homes which tend to be larger, better built and more highly equipped. A 2023 study by The Right to Build Task Force examined the Energy Performance Certificates of new self-build homes which showed that their energy consumption was between 8% and 42% less than typical new-build homes.[1] Furthermore, the homes are occupied by people who have gained satisfaction, confidence and skills in construction and management and who will look after their buildings properly. What is striking is that this community spirit has a life of its own, and this lives on even after the original self-builders have moved on and new residents move in.

People are attracted to the idea of self-build because it allows them to get at least some of the following:

- the fulfilment of a fundamental instinct to make a home, offering the satisfaction of creating something to be proud of;
- a bigger, better home for their budget, compared with developer-built houses;
- complete control over the design and specification;
- the ability to design a house with low energy demands and a low environmental impact;
- the ability to make money by turning work into equity;
- ample opportunities to acquire skills and learn about building, working with others and dealing with professionals and the authorities.

Self-help housing elsewhere

Jon covers this topic in detail in his book, *The Self-Build Book*.[2] Here, we emphasise the potential of the technology of construction to create opportunities for people to build for themselves.

The Self-Build Report UK 2021–2025 also shows that the UK has the lowest level of self-build of any developed economy, estimated to be at most 7% of the housing market (13,000 dwellings). This compares to over 50% in Austria and Germany, with other European countries around 40%, and the United States and Australia all significantly more than the UK. Jon is also familiar with Guyana and Trinidad and Tobago, where self-build is the official housing policy.

There is a self-built city in the Netherlands at Almere, where 3,000 homes have been built on plots sold by the municipality with service connections and roads. The plots come with a 'plot passport' defining the footprint and height of buildings permitted. Beyond this, pretty much anything goes. Some large plots in the Oosterwold area are available with no limitations on use or size and form of building. They are being developed for businesses, smallholdings, multi-household settlements and self-sufficient developments, all in the spirit of Dutch willingness to experiment. Elsewhere, in cities such as Den Haag, Delft and Amsterdam, there are streets of individually designed and built self-build three-storey townhouses. In Amsterdam, there are whole city blocks of apartment buildings commissioned by community groups who fitted out the individual dwellings within the shell of the contractor-built block.

BELOW The neighbourhood of Oosterwold in Almere, NL, consists of large plots for sale forming in time a self-organised city. You can build whatever you want so long as it doesn't harm others. You have to organise your own road access and services and it is envisaged that half the land will be used for urban agriculture.

The background in the UK

In the UK there has been a history of people organising their own housing in the face of the Crown, the Church and the aristocracy privatising the land over the centuries.

A century later, after the First World War, there was an agricultural recession and people were able to move out of overcrowded rented accommodation in the cities, notably London, and buy plots of abandoned land in the Thames Valley, the home counties and the south coast. They built shacks and shanties hidden way down unmade roads, without services, in what became known as the 'plotlands'. They were considered rural slums by local authorities but some remain today – many improved and extended to form desirable places to live.

There was another resurgence of self-help housing after the Second World War when many houses had been destroyed by bombing and building materials were in short supply. Areas of ordinary brick-built, houses and bungalows were built by groups of young, fit tradespeople with a mix of building skills: bricklayers, plasterers, electricians, plumbers, roofers and so on. This activity continued at a low level for a few decades but has been overtaken as housing conditions improved, land became harder to access and people became more prosperous.

The thread of people building their own homes continues, however, as the generation who benefited from post-war prosperity desire better, more individual homes than the market provides and have built large, well-equipped homes in the countryside, spurred on nowadays by TV shows such as *Grand Designs*.

There have been a number of attempts to restore land to the people. In the seventeenth century, a group of dissidents called The Diggers sought to establish a settlement on common land at St George's Hill in Surrey. This initiative failed, but the ambition came to a head again during the industrial revolution when in 1840 the Chartists devised a scheme of land settlement and five rural settlements were established. They were challenged in the courts and the settlements were sold into private ownership.

LEFT Plotland houses in Jaywick Sands on the east coast.

Segal self-build

Walter Segal designed and built a number of modest timber-framed houses in England and Ireland during the 1960s and 70s. When one of his clients, a school teacher, sent home the carpenters hired to build his house and did it himself, Segal saw an opportunity for groups building low-cost homes. This was at a time when 'community architecture' was being promoted by the then Prince Charles and the president of RIBA, Rod Hackney.

An opportunity came in Lewisham, south-east London, where the council promoted two housing developments built between 1979 and 1985. They broke new ground because the building is raised above the ground on posts, which removes the need to level the site and reduces the foundations to a limited number of isolated concrete pad foundations, which can be hand dug. There was no requirement for particular building skills or 'wet' trades such as bricklaying and plastering, merely the ability to use a saw and handheld power drill. This enabled anyone to build themselves an entire house from start to finish, radically reducing costs by up to 40%. It also allowed each household to build their own home at their own pace to fit around normal family life without having to join a team.

Accessible construction

This approach has significant advantages over the predominant organisational model for group projects, which requires individuals to commit to a fixed quota of time on site – often 20 hours a week – so can lead to difficulties with people not willing or able to fulfil their quota and suffering fines for non-performance. There can also be resentment problems arising from the fact that some people's time is more productive than others, which is difficult to assess and reflect in any rewards for self-help labour. The Lewisham model is much more flexible; households were given a timeframe for completion, within which they could work flexible hours to suit their needs. This made it easier to balance childcare, holidays and work deadlines, for example. They invited their friends and relatives to help and could employ tradespeople at their own expense, if necessary. There was also a great deal of informal collaboration where people helped one another out in the spirit of mutual support. The critical thing is that the work is completed on time; the cost of borrowing development finance was not factored into the scheme cost in a direct way at that time, so there was less pressure to complete on time than nowadays, when the cost of borrowing if a development overruns can be substantial.

In the context of a group self-build project, individuals may be offered a discount set at the value of the amount of work undertaken by the self-builder. If they do not complete work in a timely or satisfactory manner, a contractor will be bought in and the cost deducted from the self-build discount. This is a flexible arrangement that makes the self-builders responsible for progress by rewarding the value of work, not the amount of time spent on site, and provides a real incentive to complete on time.

Documentation for a Segal building

Following Walter Segal's example, we have found that preparing all the documentation, drawings, schedules of materials, building instructions and structural calculations provides control of the project without relying on consultants. This means that the designer is able to optimise the structural arrangement at the same time as developing the layout. This leads to a quick and cost-effective way of achieving an overall proposal that is well-

<u>List and Quantities of Materials for Assembly Kit</u>

Project / Address / Client: Houses at 17 Longton Ave, Sydenham, Bill Gosbee and Gordon Lewis

Materials	Description & Location	Grade	Section or Unit Size	Length & Quantity or Total Area	Finish	Price
	Quantities Listed per house!	(sizes nominal)				
Timber	Patios at rear of houses					
	beam at south end of patio	S2/50	50 x 150	1/4200	sawn	
	battens screwed to beam and facia opposite	do	50 x 50	2/4200	do	
	joists (check out over battens)	do	50 x 150	7/2100	do	
	facias to patio	Parana pine	25 x 200	2/2400 1/4200 }	PAR	
	duck boards to patio	Kerivng	25 x 150	14/4200	PAR	

LEFT Detailed schedule of materials required to build a house prepared by Walter Segal.

integrated and economical. The schedule of materials lists everything required and is used to obtain competitive quotations and an accurate figure for the cost of materials.

The use of the pattern language to discuss domestic design and options with people unfamiliar with the issues and possibilities when designing, together with the clear documentation for a Segal building and the simple and flexible construction process, all go towards helping people without previous design or construction knowledge or experience to design and build a house from start to finish. A number of architects have designed approaching 200 self-built homes in groups and one-off developments, as well as other small community buildings for various uses, based on the Segal Method.

The £28,000 house

An example of these developments is this 100 m², three-bedroom, two-bathroom house completed by its owner in 1995. It is constructed of conventional timber stud wall panels on a timber floor deck, raised above the ground with pad foundations on a sloping site, and a monopitch grass roof with deep overhangs.

This house might cost around £75,000 to build at the time of writing; perhaps a quarter of the cost of a one-off contractor-built house. More people should be aware of the reductions in cost possible when construction and information enables anyone to design and build a house using their own unpaid work, while also avoiding the profits of the housebuilders.

The limitations of Segal Method construction

Walter Segal rethought building construction from the first principles. Various people have been inspired by the simplicity and affordability of Segal's approach to develop it to modern standards, particularly with regard to energy conservation. Segal's answer was simply to wear more clothes. His approach to detailing by clamping overlapping components together allows a high degree of tolerance, which is one of the reasons these buildings are straightforward and economical to build. Contractors, too, do well with details that are not complex or require great accuracy, and such details stand up well over time and tend not to deteriorate.

The one thing that is difficult to achieve using loose-fit construction is an airtight envelope, as required by modern construction. It is also challenging to obtain high levels of insulation using common low-cost insulation materials. Construction thicknesses increase from Segal's 50mm to 300mm or more for walls, roofs and raised ground floors. What happens is that you need a secondary structure to create a thick wall. This may be in the form of spaced studs, wall trusses or 'I' studs, for example. In either case, the secondary structure becomes very strong in itself and is more than adequate to support the building without the need for a separate structural frame. The search for an energy-efficient and airtight version has led towards alternative low-energy timber-frame constructions (see Chapter 3.2).

Moving on from construction issues, Segal Method buildings often have a singular appearance: a flat roof with deep overhangs and an insistent rhythm of panels with cover battens, which many liken unfavourably to post-war prefab. This look sometimes falls foul of planning officers who want something that looks like the surrounding buildings – preferably with brick cladding. Subsequently, many designers have designed pitched roofed buildings with alternative cladding, often in timber as a cheaper alternative but one that requires regular maintenance.

Assistance for self-build

Recently, concern has been growing in some government circles about poor design, specification and construction by the monopolistic 'big six' housebuilders and the excessive profits they have made. Some MPs have been campaigning to introduce more competition in to the housing market from small developers and self-builders.

Richard Bacon MP was commissioned in 2021 by the Prime Minister to report on scaling up self-commissioned housing; legislation has been passed to require local authorities to maintain a register of people or groups wishing to self-build or commission a house from custom-build developers and to make land available for this purpose through the planning system. 'Help to Build' loans are available and such development is free of Community Infrastructure Levy (CIL).

The barriers to self-build

It is not only the complexity of modern construction that limits opportunities for self-build but also the increasing complexity of the regime of regulation, with over-complicated planning law, unreasonable avoidance of risk (see Chapter 4.6) and sensible improvements in safety and performance.

In the UK, access to land (see Chapter 4.1) and finance are significant barriers, as is the necessary infrastructure of enabling professionals with experience of dealing with inexpert clients (see Chapter 1.6). Finally, in the UK there is a cultural prejudice among housing officers, financiers, lawyers, designers and others, which suggests that ordinary people are not to be trusted to make sensible decisions or to deal with significant sums of money.

Self-build opportunities

While building or commissioning your own home individually or as a member of a group is not for everyone – most people just want an adequate dwelling at a reasonable price – the opportunity to do so should be made more available and a necessary complement of any satisfactory public or private mass housing policy.

A self-build story that came right after tribulations along the way.

The opportunity was a windfall site occupied by under-used lock-up garages which was acquired by a large housing association when it purchased the surrounding estate. The now defunct Community Self-Build Agency founded in 1989 provided initial advice and support and Jon worked with a group of local residents to design a low-energy development of 10 two- three- and four-bedroom houses.

Grant and loan funding was provided as part of the association's programme of new homes for social rent. Project managers were appointed by the association who took the lead and who were concerned about the perceived risk of self-build and who decided to limit the self-build element to finishing shells constructed by a design and build contractor. The contractor purported to have experience of self-build but this turned out to be very limited.

SELF-BUILD VS CUSTOM BUILD

--

Self-build and custom build are 'self-commissioned' homes to the owner's design and specification. Self-builders have a choice over who builds the property, or they can do it themselves. Custom builders have no choice of builder and have somewhat more limited choices over design and specification.

Cuts to the design and specification were made to reduce costs and the project managers acted as Employers Agent but the build quality was poor which prejudiced the energy performance. Members of the group undertook training at the local technical college and fixed some fittings and carried out decorations.

The residents had moved from often overcrowded, damp and expensive accommodation and were overjoyed with their new, comfortable homes on low rents, and they gained skills, self-confidence and good neighbours along the way.

Overall it took nine years from the first idea to completion which is unreasonably long for people to work for a decent, affordable place to live but not unusual for community self-build projects.

LEFT Headway Gardens self-build Walthamstow, design by Jon Broome Architects 2008.

BELOW A mother-of-three had been on the waiting list for 12 years and had shared a bedroom with one of her children for 18 years. She learned skills such as kitchen fitting. 'It wasn't just doing your own work in your own home, it was a group project that we helped each other out,' she said.

Sustainable neighbourhoods

One of the fundamentals of sustainable systems is that in order to be sustainable in the long term they must have support from the people affected. They must include users in the design of the system, its implementation and long-term maintenance. A sustainable living environment relies on creating a planning system that allows people to participate in the major decisions involved (see Chapter 4.6 for more on the planning system). David Rudlin and Nicholas Falk expand on the history and background of the development of British cities and city planning in *Sustainable Urban Neighbourhood: Building the 21st century home*.[1] Here we concentrate on the role of community-led housing.

An ideal of sustainable living is the creation of so-called self-sufficient dwellings set within a smallholding, which can provide some of their food, energy and water and recycle waste. This is only possible at low densities. Self-sufficient settlements can, however, support a relatively small population and there are few examples of sustainable quarters in towns or cities. What is required in an urban setting is to radically reduce the consumption of resources and the generation of waste. London draws resources from all parts of the world, creates pollution that affects a wide area and sends waste out to sea. Its population of 8 million was estimated in Herbert Girardet's 1994 article 'From Mobilisation to Civilisation'[2] to consume every year more than a billion tonnes of water, 20 million tonnes of fuel oil and 8 million tonnes of building materials, while creating 7.5 million tonnes of sewage sludge and emitting 60 million tonnes of CO_2.

It is tempting to conclude that cities are an environmental disaster. However, while in less-developed economies subsistence in rural areas has a low environmental impact, this is not so in the developed economies where people living in the country expect the same level of activity and services as city dwellers (but with fewer coffee shops, perhaps). Getting access to those services in rural areas leads to around twice as much car use as well as increased impacts from the distribution of goods and services and additional infrastructure to provide water, power and broadband, for example. This is reflected in council tax rates, which tend to be higher in rural areas. Abandoning towns and cities for the dream of rural self-sufficiency is not a solution, but neither is a depopulated countryside with a few key workers to manage factory farms.

Transforming the urban economy

It would seem that the dense, walkable city is a much more environmentally efficient form of settlement while the prevalent form of suburban development is environmentally undesirable. 'Eco-villages' on greenfield sites are not the way to go; however, the urban economy needs to be fundamentally transformed to deliver a sustainable future. We need to move towards a circular economy that uses the outputs of waste as the resource to create what we need. Resource inputs of energy for heating, power and water can be reduced to minimum levels, much of which can be provided by the building itself. Local rather than imported resources can be used, food waste composted and food grown on roofs and in gardens. Much waste is exported to the Far East, where some is illegally dumped or burnt. More waste can be reused and recycled; in energy from waste plants, in schemes both imposing deposits on reusable bottles and taking back used appliances for new. Cities create opportunities for new ecological businesses; recycling metals, plastics and glass, for example. The city of Curitiba in Brazil has established nurseries linked to recycling centres which are the source of play equipment, as well as colleges equipped with computers taken from the waste stream.

BELOW Vauban neighbourhood cafe in Freiburg, Germany, in a development by 200 housing co-ops with varied low-energy homes around green open spaces and served by a new tram line constructed at the outset.

The rural economy will need to change. Commuter dormitories are unsustainable and the self-sufficient 'good life' is a lot harder than it looks. We suspect that a more sustainable rural settlement will need to emerge organically rather than be imposed as a concept by urban designers or politicians. Nick has spent most of his life living in the countryside. Intellectually he supports sustainable urban development over a move out of towns and cities, but he knows that city life does not suit everyone.

Traffic

What would a sustainable neighbourhood be like? First, there would be very few cars; walking and cycling would be the norm for local journeys. Layouts should be easy to negotiate as a pedestrian or cyclist – not confined to subways and overbridges – with streets that connect rather than a maze of cul-de-sacs. Streets would be the focus of activity, lined with buildings with active frontages and plenty of people walking about. This would, in turn, promote security rather than deserted pedestrianised routes between 'defensible' spaces.

Creating pedestrian-friendly places does not mean that cars are banned; rather than they do not have priority. Vauban in Freiburg, South Germany, is a new neighbourhood of 5,000 residents in which cars can use the network of narrow roads to drop off and pick up but not to stay longer than 20 minutes. The roads have no segregated pavements and are populated by pedestrians and children most of the time.

The Dutch have play streets where cars mix with children. In Amsterdam, the streets are not pedestrianised but they are narrow, lined by big trees, paved with uneven cobbles, and shared with crowds of people and legions of bicycles. Humpback bridges limit visibility, motor traffic moves at no more than 15mph and residents put their tables, chairs, potted plants and sofas out in the street on a sunny day in the heart of the city to enjoy a meal while watching the world go by.

This brings to mind Jon's dad who, 60 years ago, used to say that what is needed is road-worsening schemes not improvements. Improvements make it easier to drive so more people travel by car until the roads become congested again. The point is that it's no good just reducing traffic alone, you have to simultaneously reduce the capacity of the system to just below the level at which it becomes saturated, by introducing bus lanes or limiting parking, for example.

A neighbourhood with radically reduced traffic requires good public transport. The economics of transport require sufficient density to support a viable bus or tram service. A walkable neighbourhood also requires that services, shops, schools and workplaces are within walking distance, which also needs relatively dense mixed-use development.

Open spaces

Density is also affected by the amount of open space. There is an assumption that green space in built-up areas is a good thing, which is not wrong, though it is rarely the whole story. Lots of open space is a real problem; either it is deserted, particularly at night, and therefore potentially dangerous, or it may be space leftover after planning; a desert of

ABOVE Cyclists and pedestrians have safe space on this street in Leeuwarden, NL, with wildflower planting.

grass without any clear function (beside a busy road, for instance). It needs cutting and is devoid of the biodiversity that can make natural areas so useful in cities, with trees to reduce pollution, provide shelter and shade in summer and become havens for wildlife, some of which may be threatened by monocultural farming in the countryside. Open space is necessary for sport, play, growing food, meeting people and watching the world go by, but it must be carefully designed with a clear purpose in mind and not neglected.

Nan Fairbrother remarked in *The Nature of Landscape Design* that architecture is concerned with the form of buildings and spaces within.[3] The spaces in between, however, are what we appreciate when we move around towns and cities and it is the shape and scale of the urban landscape which defines the character of our environment.

Energy

The form of buildings – height and proportion, flats or houses, terraced or detached – has a fundamental influence on their energy performance, as we examine in Chapter 2.3. Low-energy buildings to Passivhaus standards or equivalent dramatically reduce energy consumption, which in turn reduces emissions and running costs, making homes more affordable in the long run. This improves comfort and eliminates the risk of condensation and mould growth, thus improving health and well-being and reducing demand on the health service.

Mixed uses

Neighbourhoods need to incorporate a range of uses, including local shops, cafes, meeting places and workspaces. They should be integrated into the fabric of the city, unlike the council or private estates as we know them, or even gated developments.

Dwellings would be available in a range of sizes to reflect the local population, including:

- shared dwellings suitable for single people;
- small flats for couples;
- maisonettes and houses for families;
- lodgings designed to make life easy for infirm and elderly people (and everyone else).

They would be offered in a range of tenures to reflect different life stages and incomes locally, including rents with subsidy, shared ownership, outright sales and collective ownership through a co-operative or Mutual Home Ownership arrangement[4] where instead of owning an individual property, or a percentage share of the value of an individual property, residents own equity shares in a mutual property trust owned by them and other residents.

Dwellings need to be adaptable to ensure longevity. This can be enhanced by building to greater than minimum floor area - the UK currently has among the lowest space standards in Europe. More space allows different uses and internal arrangements, as needs and wishes change over the years. We discuss adaptability in more detail in Chapter 1.8.

The role of residents

To return to the opening assertion, for a system to be sustainable it has to include users in its design, implementation and long-term maintenance; in other words, residents have to be involved in the major decisions that determine the design, construction and management of their homes. Not everyone wants to spend the time; some preferring to get on with life in other ways. Nevertheless, many feel alienated by their environment and more still are physically and mentally disadvantaged by their living circumstances.

A housing system needs alternatives to take-it-or-leave-it mainstream mass housing for it to be successful. Participation in housing has been more common in Northern Europe and Scandinavia and an infrastructure of professional enablers exists to bring banks, designers and local authorities together with groups of residents. The benefits of this approach were recognised in the UK, and community architecture became a movement in the mid-1970s. A variety of projects were realised over the following decades, such as community self-build projects in South London and Bristol, the refurbishment of Victorian terraced houses in Macclesfield and other towns and cities, and new-build housing co-operatives in Liverpool, in particular. The impetus was lost in the 1990s as governments became preoccupied with other issues, value for money and consolidation of the housing association sector as developers who had taken over responsibility from the local authorities for housing for people on low incomes.

Multigenerational housing at Melfield Gardens, Lewisham. Levitt Bernstein 2025. 30 affordable homes for residents aged 55 and above as well as two four-bedroom homes for eight postgraduate students from a local university. Each student will spend a number of hours assisting older residents by offering company or participating in the cultural and recreational activities that take place in the shared south-facing 'garden room'.

Co-ops, co-housing and community land trusts

A couple of co-housing developments followed later in the 2000s in Stroud and Lancaster. Co-housing was pioneered in Denmark in the 1970s and is now popular in the Netherlands and the United States, among other places. It is based on the idea of a group wishing to combine communal living with household privacy. There are a significant number of communal facilities for cooking and dining, nurseries, playrooms and workspaces, while residents live in individual dwellings with their household, unlike a commune with an entirely shared living space. The dwellings are often smaller than normal, taking account of the communal space outside the home.

Another recent development has been the establishment of community land trusts (CLTs) on a model pioneered in the southern states of America around the same time. These are set up and run by people (not solely prospective residents) in a community to develop and manage not just homes but also other assets important to that community, like community enterprises, food growing or workspaces. CLTs act as long-term stewards of land, ensuring that what is built on it remains genuinely affordable, based on what people actually earn in their area, not just for now but for every future occupier.

Housing co-operatives

Another model for community involvement in housing is the housing co-operative. These were established in Germany at the end of the nineteenth century and are common in Northern Europe and Scandinavia. Many were established in the UK during the 1970s. They have a limitation; most are 'fully mutual', which means that all members are residents and all residents are members.

CLTs, together with co-housing and housing co-operatives, all share three common principles:

1. The community is integrally involved throughout the process in key decisions, such as what is provided, where and for whom. They don't necessarily have to initiate the conversation or build the homes themselves.
2. There is a presumption that the community group will take a long-term formal role in the ownership, stewardship and management of homes.
3. The benefits of the scheme to the local area or specified community group are clearly defined and legally protected in perpetuity.

Partnerships

Community-led housing projects can be initiated by a small group coming together independently to:

- develop an idea;

- find a site;
- prepare a feasibility study and business plan;
- obtain finance for the planning stages and construction;
- develop a design and obtain planning permission;
- recruit residents;
- prepare detailed construction information;
- oversee the building works;
- organise long-term finance, maintenance and management.

This is a lot of work but it does ensure that the people involved have full control over the decisions required about the development and how it is funded.

Alternatively, many groups form a partnership with a housing association or supportive developer. This brings assets as collateral to support loans, expertise in project control, finance and housing management. An organisation with experience and track record also offers credibility to lenders, the local authority and others. It does, however, tend to mean that residents may have less control over the decisions being taken on their behalf because it will often be an unequal partnership within which the dominant housing association will be making the critical decisions.

Obstacles

Community-led housing in the UK has been held back by three principal issues. First, the difficulty accessing building land – the background of this is discussed in Chapter 4.1, where we outline one policy approach employed by municipal governments in Germany and elsewhere to make land available for community-led housing. It is not so straightforward in the UK because local authorities had the ability to acquire land at 'existing use' prices backed by compulsory purchase orders removed by the Westminster government in 1961 and because local authorities are under government austerity measures to maximise income from land and other disposals. Nevertheless, some authorities have been able to dispose of land for no charge under the Public Services (Social Value) Act 2012 on the basis of the social benefits that accrue from the proposals envisaged.

Second, the UK does not yet have an established network of organisations and advisers to facilitate community-led initiatives; to provide advice and support, seed money to get started, and experience and credibility to navigate the world of housing development (which is complex to the uninitiated). A network of enablers is currently being established by the umbrella group, Community Led Homes. Commercial developers have assets, which they can quickly mobilise to raise money and are difficult for community groups to compete against.

Finally, raising finance for social projects is not easy in the UK. Not-for-profit housing organisations in Germany and elsewhere on the continent enjoy operating in an economy that still has a regional banking infrastructure, unlike in the UK, where the Midland Bank, the Yorkshire Bank and others were taken over years ago. The regional banks on the continent are familiar with the aims and track record of local authorities, community organisations and their advisers. They view self-help housing groups as good lending risks; they cannot afford to fail. This is in contrast to the situation in the UK, where community and self-help housing is viewed as risky business by the UK's global banks, who can make more money out of gambling rather than lending. This leaves a few social sector lenders who are prepared to lend, but they are wary of lending scarce resources to housing groups without imposing stringent conditions and relatively high interest rates.

In the UK, current government limits on local authority spending mean that councils are generally unable to provide feasibility, development or long-term mortgage finance as they have in the past. Financial support from government has become recently available in the UK for community-led housing in the form of capital and revenue grants. These grants have not been continued and community-led housing is now funded through the normal affordable housing fund. However, government grants are constrained by complex rules (the prospectus runs to over 50 pages and the guidance for making an application is an additional 240 pages).

If you aim to own homes for rent to low-income households – 'social housing' – then there is in addition an even more involved application process to become a 'registered provider', which normally takes around 12 months to complete. Some of the aims of community-led housing – notably the aim of creating genuinely affordable homes and retaining them as community assets in perpetuity – are in conflict with the government policy of increasing the market for individual home ownership, whereby residents ultimately have the right to sell their homes on the open market, which renders them unaffordable for subsequent residents.

RIGHT 11 one and two bedroom flats designed with residents and neighbours by Archio for discounted sales by London CLT 2022.

Funding community-led housing

You will need initial start-up loans and grants for feasibility studies, a business plan, precontract funding to develop the scheme, a development loan for construction that will be repaid by sales income, and a long-term mortgage on completion. Together with management and maintenance costs, this will be repaid by rental income, which in turn will set the level of affordability for residents.

In a project to create a sustainable community, you would aim to make homes available at a cost that is affordable within the range of income of the local population. This leads to the question of what proportion of income is reasonable to spend on housing costs.

Achieving the 30% target for households with average earnings is not easy and puts pressure on standards, so new dwellings tend to be small with relatively poor energy performance. This increases pressure on household incomes due to higher heating costs. Small dwellings tend to cost relatively more to build per square metre of floor space than large ones because both require similar electrical, plumbing, heating, kitchen and bathroom installations. However, this is set against an inverse ratio of value to cost, which values small flats at a higher value per unit of cost than large houses. This means that small dwellings are more 'profitable' than large ones and explains why commercial developers try to build a high proportion of small dwellings and relatively few large ones.

The type of tenure is also critical because outright market sales realise more income than social rent, for example. This means that a balance has to be struck between genuinely affordable homes and space and energy standards, as well as the proportion of affordable homes and the mix of large or small dwellings in a scheme. There are, as you can see, a lot of moving parts based on a range of assumptions that have to be manipulated in a financial model to arrive at a balanced solution. The assumptions will almost certainly have to be adjusted during the development of the project.

AFFORDABILITY

--

Housing costs should (but often do not) include not only rent or mortgage repayments but also electricity, gas, water, phone and internet charges as well as council tax. It has been considered that housing costs should be no more than 30% of household income, although current market conditions mean that many pay at least 50% and some up to 75%. Meanwhile, the government considers that paying 80% of market cost of buying or renting is affordable, although a significant proportion of the population cannot afford this.

Towards a sustainable neighbourhood

Jon has been involved with one CLT, Rural Urban Synthesis Society (RUSS) in Lewisham, South London. RUSS has overcome these obstacles to build a sustainable neighbourhood of 36 dwellings with a shared guest room, office, laundry and meeting room with kitchen.

The project was inspired by a son of one of the original Lewisham Self-Builders who completed their family house in 1985. Kareem Dayes was looking for an affordable place to live in London and he set up RUSS in 2009 with ambitious aspirations for sustainability. He was aware that the Segal self-build houses that were occupied on shared ownership leasehold terms had all reverted to the open market and thus become unaffordable to young people starting out. For this reason, resale covenants have been put in place which limit any increase in prices. The aim was to see what a sustainable neighbourhood could be like, as there were no direct comparisons in the UK.

The numbers of one-, two-, three- and four-bedroom homes reflect the household composition of the local community and are available in a range of tenures: social rent, discounted affordable rent for shared flats, shared ownership and 80% fixed equity sales. Housing costs reflect local incomes but at a target 20% reduction from market rates. The homes were designed 10% above minimum space standards to enhance adaptability. They are well insulated, properly ventilated, airtight and thermal bridge-free to reduce energy costs and thus enhance affordability and reduce emissions.

Potential residents held workshops, which decided on 10 principles to guide the project. These have been applied to the design, construction and management of the project from the start, and are as follows:

1. RUSS will create socially, environmentally and economically sustainable neighbourhoods in the city.
2. Neighbourhoods should balance the interests of residents, the wider community and the council as landowner.
3. RUSS should build truly affordable homes.

BELOW The dwellings at RUSS all have a south-facing open space outside the front door on the access deck.

BOTTOM RUSS Community Land Trust, Church Grove, Lewisham 2024. Project strategic design by Jon Broome Architects, co-design and scheme design by Architype and detailed design by SEH Architects.

4. Decisions that affect the neighbourhoods should be under the control of residents.
5. Developments should be embedded in the local community and include space for community use.
6. Neighbourhoods should reflect the local population with a mix of families, couples and single people, both young and old, and a range of incomes.
7. RUSS neighbourhoods should not only reduce environmental impacts by efficiently using energy and building materials, but should proactively create resources of power, water and food.
8. Residents should have the opportunity to be involved in the design, construction and management of neighbourhoods.
9. RUSS developments should create opportunities for training in organising and building for residents and others.
10. Our projects should be self-financing with robust financing and delivery systems.

Some of the flats have been fitted out internally by the occupants, who have designed the layout and specification of their homes while acquiring skills. The project is an ambitious one and has had a long gestation due to the following obstacles:

- The local authority was of the view that they could complete the project better than a voluntary organisation, so they held a public meeting but then did nothing further.
- RUSS identified a site and the council made it available at no cost but required a long and costly competitive process to secure the land.
- Finances were difficult to negotiate with banks, building societies and grant-giving authorities who were unfamiliar with the idea of community-led housing.
- Planning permission took a year to obtain with no less than 37 conditions attached.

Some compromises have been necessary because of problems with the budget; principally reductions in space and energy standards and omission of access to the roof for food growing, as well as a concrete frame instead of a timber one to avoid the perception of risk, following the Grenfell disaster. The shared meeting room and kitchen was omitted to save money and so the residents fundraised, designed and self-built the 'Hub', which is now used by the RUSS School of Community Housing for hosting educational seminars on community-led housing and by local community groups.

At the time of writing, the homes are just being occupied, an inconceivable 15 years after the first inspiration, and we look forward to a new neighbourhood of self-confident, self-reliant residents in affordable, low-energy homes. Experience has shown that communities of this kind improve the surrounding areas and the general health, education and employment outcomes of residents. Governments should be more aware of the economic, social and environmental benefits that can follow. It's tragic that the benefits of the original self-build projects had been lost over the intervening 40 years by the local authority that had no institutional memory of what had been achieved when 'community architecture' was a 'thing' in the 1970s and 80s.

ABOVE The inside and outside of the RUSS community hub, fundraised, designed, constructed and managed by the residents. The project also provides a guest suite, shared laundry and office space.

An ex-military base, the Vauban District of Freiburg has been labelled the most sustainable town in Europe. As in this image, the buildings mostly provide a quiet supporting role for people and nature.

2.1 Comfort and sufficiency Both important considerations, but what kind of 'comfort' and how much space, complexity and technology do we need?

2.2 Closing the performance gap Why many buildings do not achieve their planned minimum-emission performance

2.3 Form factor, massing and shape How size and shape are often fixed before detailed cost, energy or structural analysis are performed

2.4 Environmental modelling and targets Numbers and targets are crucial to an understanding of how to improve the performance of buildings

2.5 Passive solar design Shifting the focus from passive solar gains to glazing design to serve the comfort of occupants

2.6 Why Passivhaus? The benefits, limitations and myths surrounding the use of this highly effective building standard

2.7 Buildings must breathe Draught-free construction – and the resulting need for moisture and odour control – are essential

Comfort and sufficiency

Many factors influence comfort, but in terms of building we are mostly worried about thermal comfort. Simplistically, we think of comfort relating to air temperature and perhaps at the extremes, humidity, but it turns out to be more complicated. Research in Denmark in the 1970s by Povl Ole Fanger and his colleagues quantified a number of factors that affect human thermal comfort.[1] They found that in heated buildings, humans are very sensitive to draughts, temperature asymmetry (in this case the difference in radiant temperature experienced by opposite sides of the body) and the temperature difference between the head and feet. This research allows us to predict the percentage of people who are likely to feel comfortable for a given combination of these measurable factors.

Clearly, we all have different preferences and sensitivities, and these will change as we age, but this approach allows us to set conditions in non-domestic buildings that will be acceptable for the majority of people. We find, for example, that large areas of glazing or quite gentle draughts require an increase in air temperature to compensate. This in turn leads to increased energy use to maintain this higher temperature. Understanding the principles is also useful for diagnosing why individual people feel cold and why it may not be just a case of adjusting the thermostat.

Applying the lessons, we realise that reducing draughts and radiant asymmetry means that a wider range of people will be comfortable with the same air temperature. This is great in a care home, say, where elderly and sedentary people, sensitive to draughts and cold ankles, are living alongside young people carrying out heavy lifting and other energetic activities. The elimination of cold draughts allows the same comfort at a lower temperature. More people will be comfortable, despite the large range in age and physical activity.

Edwards Court Extra Care Housing Scheme in Exeter puts this theory to practice.[2] Although occupants can choose to open the window or increase the temperature of their flats, the building tends to sit at about 21°C, which suits the older occupants and the younger staff alike. As Claire Taylor, property services manager says, 'We have an age range of 50 to 101 and to never get, even in the depth of winter, any comments about the heating is a major compliment!'

ABOVE This photo of Old Holloway by architect and self-builder Juraj Mikurčík has been used widely to try and convey the comfort experienced in Passivhaus buildings.

Comfortable buildings

While there is ongoing research and debate about the details and equations, we might assume that a good building is comfortable. At this point opinion divides, and as with so many of the seemingly objective topics covered in this book, emotions can run high. There is a school of thought that argues that this pursuit of comfort is a truly dystopian route to a grey neutrality devoid of thermal delight. According to Richard de Dear:[3]

> *'Lisa Heschong's book* Thermal Delight in Architecture[4] *succinctly put forward the case that architecture was profoundly impoverished when it outsourced the responsibility for the thermal realm of buildings to engineers. Heschong contends that, under certain combinations and sequences, the elements of indoor climate can infuse our total sensory experience of space with layers of affect, emotion and even delight, in ways that other dimensions of the built environment cannot. But these opportunities are squandered when thermal design falls into the hands of those whose stated mission is to neutralise buildings.'*

But is the problem that people are too comfortable or that buildings have become over-reliant on mechanical air conditioning? In the UK it is mostly offices and other commercial buildings that are air-conditioned, but in hotter climates air conditioning (not to be confused with simple cooling or heating) is more common. We believe that the building should do most of the work to deliver comfort, and thermal comfort should not be outsourced to engineers. Some twist this argument to excuse buildings that are uncomfortable by design. If, for example, thermal discomfort caused by a large area of cold glazing is experienced, this can inform the approach taken when designing a building in a number of ways. The building occupants can be convinced that it is a dynamic feature of a lively building. If unconvinced, the engineer could be asked for a solution such as installing trench heaters under the windows, though this is expensive and won't help in the summer when the glass gets warm. High-performance triple glazing can also be specified, which again would only help in winter and the area of the glazing could be reduced.

Finally, although a number of academics have got into heated arguments when discussing comfort, no client or building occupant has, to our knowledge, ever complained of being too comfortable in a building. We won't judge these academics for masochistic tastes but we don't think they should be imposed on the occupants of buildings. A 'total sensory experience' of space with layers of affect, emotion, even delight, is not how we would describe a cold, damp flat with mould spores in the air.

Designing for comfort is necessary; but not everyone's idea of comfort is the same. Easy access to modify the thermal environment is important – windows that can easily be opened and adjustable thermostats, for example – but the imposition of negative factors which are not easily ameliorated, such as large areas of glazing causing cold downdraughts in winter and overheating in summer, are to be avoided.

Sufficiency trumps efficiency

Efficiency may be described as doing more with less, which we would imagine is something most would support: buildings that stay warm with less heat, thanks to efficient heating systems and insulation; ventilation with plenty of fresh air all the time but without cold draughts and minimal heat loss, thanks to efficient heat recovery; energy-efficient windows that are warm enough to sit next to when the temperature plummets below freezing outside.

As with comfort, not everyone gets excited about efficiency. Some will worry about upfront cost, others about loss of architectural purity, yearning for single-glazed thin steel frames and slender cantilevered concrete without the layer of fat that insulation inevitably adds. Others appear to take a more puritanical view and suggest that efficiency makes us soft, hardship is character building, and we should just wrap up warm and throw the windows open – arguably the most efficient option. Here, the argument is that pain is good for us, which is a variation on the 'pain is pleasure' argument previously explored.

BELOW Architect Dominic Stevens took his cue from the tradition of the 'tiny house' in Ireland when he designed and built this three bedspace 60m² house built in 50 days for €25,000 in 2010. It is well insulated and is heated by a 1.5kW electric fire.

If we are serious about the climate emergency or any of the other threats to the environment that humans depend on, efficiency allows us to reduce our use of resources to satisfy the perceived needs of the majority without undue sacrifice or coercion. However, efficiency is only part of the equation, as we also need to address sufficiency. Indeed, without a limit on how much we consume, efficiency is likely to increase consumption. This was identified by William Stanley Jevons in 1865 when dramatic improvements in steam engine efficiency led to a huge increase in the use of coal.

There are plenty of examples of large energy-efficient homes that use less heat than much smaller homes. Efficiencies of design and construction also keep the cost down, but a bigger building uses more land and materials, and will be filled with more resource-intensive stuff.

Less is more

Efficiency may be the cause in part, but it is affluence that really drives this. Making everything less efficient is not the answer. Rather than doing more with less, we need to do less with much less. This is not an easy sell for ourselves and it is taboo to suggest to a client asking for a shiny new building or an extension.

This problem was neatly surmised in an image posted to social media by Australian architect Jennifer Crawford in 2022 (see image right).

So how do we embrace sufficiency? The building regulations say nothing, and planning puts up land prices so that the rational thing is to maximise plot value with a big building. Detractors of our benchmark for building efficiency, Passivhaus, point to the resources needed to build a large detached Passivhaus home for a retired couple in a rural area. They even argue that the energy (upfront or operational) per square metre of useful floor area metric encourages larger buildings.

We do not advocate rules or even targets for maximum space for human habitation, but we can start thinking in terms of energy, carbon or capital cost per person, whether it is housing, schools or offices. Limits would be very difficult to adopt, even as a voluntary environmental standard, as occupancy is unlikely to correlate with the number of 'bedrooms' labelled on the plans.

Even without the stick of regulations, designers and planners can think in terms of how a building occupancy might change over time. As kids leave home, can the building be divided for rental for strangers, friends or elderly relatives?

Wrestling with how to reduce the impact of his Toronto home, architect and writer Lloyd Alter realised that splitting his home in half and letting his daughter's family move back in would have an immediate environmental benefit. His book *Living the 1.5 Degree Lifestyle*[5] highlights the importance of sufficiency in tackling the climate emergency. Efficient buildings are not the problem but they are also not the whole solution.

As with efficiency, we can approach this from a dry, factual angle in terms of budgets for carbon, water, land or minerals. We can take an ethical view and embrace austerity as a virtue or we might get genuinely excited about embracing a simpler lifestyle. However, it is easier to enthuse about a new eco-building or an electric car with all the latest technology.

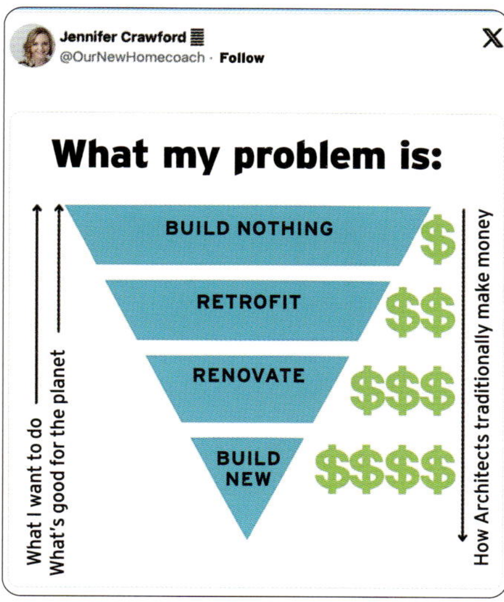

ABOVE Posted on an ex social media site, Australian architect Jennifer Crawford highlights the conflict between minimising impact on the planet and making money as a building professional. Based on the Institute of Structural Engineers' Hierarchy of Net Zero Design.

ABOVE Craft valued for something other than money at a Buddhist temple in Japan.

A case and a place for inefficiency

Reflecting on the question implied at the start, would anyone argue for inefficiency? Mahatma Gandhi promoted homespun cloth as a protest against imported industrial cloth that he saw stealing people's livelihood. Until quite recently, politicians of all Indian parties would publicly wear the coarse hand-spun and woven khadi while the general population embraced the shiniest of industrial fabrics. Industrial production is clearly more efficient in terms of labour, so it will cost jobs, but even Gandhi's famous protest had little more than symbolic impact.

It is interesting to think that inefficiency is a prerequisite for most hobbies and leisure activities; something we have all been promised we will have more time for, since the 1960s, thanks to industrial efficiency. If you want to feed your family you don't spend hours tying flies then pay a considerable sum for fishing rights on a salmon river hundreds of miles away. Makers of pots, baskets or furniture wanting to make a living find they can only do this if their name commands a price premium, usually post mortem as the studio potter Mick Casson once commented. Otherwise other income, perhaps teaching their craft to hobbyists, pays the bills. The hobby potter could embrace industrial efficiency by using slip casting and other production techniques but is more likely to seek out less efficient options such as digging and processing their own clay and hand-building with coils.

Indeed, it could be the pursuit of deeply inefficient but environmentally benign leisure activities that allow us to curtail our appetite for consumption. In the context of this book, the self-builder or person on short-term rent can take pleasure in making furniture and fittings, rather than ordering flat-pack furniture online. This might involve a furniture-making course at the local college or skip-diving for old bricks and broken floorboards to fashion bookshelves. Friends can exchange skills, labour and spare materials. This takes considerable effort compared to click-now-pay-later internet purchases, but could be just the excuse we need for social contact without having to sign up for dance lessons.

As designers of buildings of whatever type, there will be chances to design in desirable inefficiencies. Places for people to meet and chat, private or shared workshop space or gardens that rely on the hit-and-miss input of residents, rather than a contractor rolling out efficient, low-maintenance plantings with all the charm of a supermarket car park, or for more affluent developments, the precious untouchability of an overdesigned landscape or award-winning concept garden.

'The two things that really drew me to vinyl were the expense and the inconvenience.'

Closing the performance gap

Nineteenth-century industrialist Benjamin Brewster[1] first said, 'In theory, there is no difference between theory and practice, while in practice, there is.' As a bare minimum, we might hope that any building – never mind our ever-elusive good building – should do what it says on the tin, though this is rarely the case. While any product or service might occasionally fall short of what is promised, it is only in the field of building that very significant underperformance has been accepted as the norm.

To simplify the discussion we will for now put aside harder-to-measure, subjective claims such as beauty, inspiration and community enhancement and consider more easily quantifiable and predictable design goals such as fire safety, energy consumption, comfort and air quality. For example, a study by the Usable Building Trust of 23 non-domestic buildings that had been described as exemplary were found to use on average twice as much energy as predicted by design. Other studies by the Carbon Trust have identified buildings using five times as much energy as predicted and there are non-domestic buildings using up to 10 times as much energy as predicted by design. It is generally accepted that we are talking about a significant gap that would not be tolerated in other industries.

It was this gap that inspired the physicist Dr Wolfgang Feist, founder of the Passivhaus Institut (PHI), to get to the bottom of the problem.

Meanwhile, in the UK, monitoring of a development of low-energy homes at Stamford Brook by researchers at Leeds Metropolitan University identified the significant – but at the time surprising – issue of thermal bypass in party walls. Thermal bypass refers to the reduction in the performance of insulation caused by air moving through it. It is a significant contributor to the performance gap in terms of heating demand. The architect Mark Siddall[2] has pulled together lots of research into one comprehensive review.

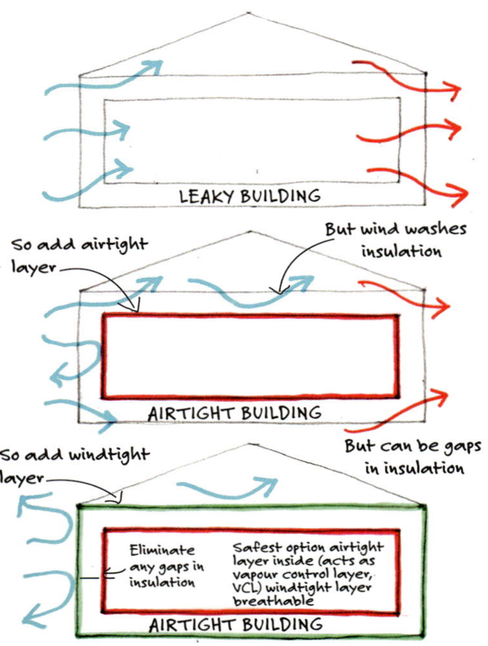

ABOVE To work as intended, insulation must be installed without gaps or cavities between an airtight and windtight layer.

BELOW Thermal image of an insulated ceiling in a low energy, housing association Code 4, home showing cooler (blue) areas caused by wind washing.

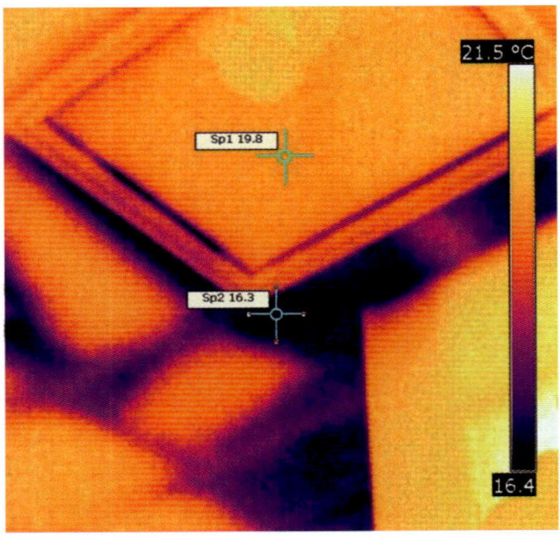

Thermal bypass

Simply put, the gaps between and behind insulation allow the movement of air, driven by wind and convection. The European Standard ISO 6946:2007 includes a table of so-called corrections and an equation to allow for voids and gaps. We know that most designers or insulation suppliers either do not know of this correction or simply choose to assume there will be no gaps. However, unlike thermal bridges, we can't calculate thermal bypass as there are too many variables; instead, we need to simply design it out. The impact depends on many variables but we are looking at U-values being up to five times worse than expected rather than 10 per cent – this is huge. To get a feel for this, imagine trying to stay warm in a freezing wind. Would you choose one woolly sweater and a thin but windtight jacket or two thick sweaters? Our leaky buildings don't even have the benefit of our airtight skin, so the wind might blow straight through the building, displacing the warm air with it. Adding an airtight layer on the warm side of the insulation – primarily as a vapour control layer to prevent interstitial condensation – stops this direct air exchange but the wind can still wash through the insulation, as with our thick sweaters in the freezing gale. To solve this, we need a wind barrier on the outside of the insulation – an outer jacket.

We might imagine the problem of thermal bypass by air movement is now solved. However, if there are any gaps in the insulation then air in the gaps on the warm side of even a fully encapsulated wall construction will heat up, and becoming less dense, rise up. On the cold side any air will cool down and sink, creating convection within the wall assembly. This will transfer heat from inside to out, bypassing the insulation.

While all construction types could suffer from thermal bypass, it is more likely when the insulation is provided by rigid boards in partially filled cavities or slotted between studs or rafters. This is a popular approach in the UK, as in theory the use of foam insulation boards with half the conductivity of air-based fluffy insulation allows walls to achieve the required U-values with nearly half the insulation thickness.

Unfortunately, it is pretty much impossible to avoid gaps behind and between these boards. This then leads to air movement bypassing the insulation. Back to our thought experiment, imagine the impossible task of making a functional insulated jacket from bits of rigid board tied together with mechanical fixings.

To prevent thermal bypass we need good airtightness on the warm side, a continuous windtight layer on the outside and no gaps between or behind the insulation.

Thermal bridges

Staying with the thermal envelope, thermal bridges are another contributor to the performance gap. As insulation got thicker, the heat loss through timber wall studs or cavity ties became more significant. Even timber studs are about three to six times more conductive than the insulation between them. Standard U-value calculations include these repeating thermal bridges but the assumptions for timber fractions, for example, are usually very optimistic so the impact is underestimated.

Other thermal bridges include structure and fixings that pass through the insulation layer. Examples include balconies and canopies. Installing windows in line with the rainscreen layer introduces a significant thermal bridge that we will look at in Chapter 4.4. When we first started calculating thermal bridges we were surprised to find they added around 30% to the calculated heating energy for a typical house.

Other causes

The people at the PHI identified these and other reasons for the performance gap between design and reality. Good airtightness combined with mechanical ventilation (with or without heat recovery) allows us to calculate the heat loss due to outside air exchange in a way that is not possible if we rely on weather-dependent draughts and random opening of windows to ventilate in winter. As we move to very low-energy buildings, the assumptions we use in our energy models for free-heat gains matter more as this can meet around one-third of our heating demand, even with efficient appliances and lighting.

It is a design problem

Thermal bypass could be blamed on construction quality but it is primarily a design problem. Fitting rigid insulation boards into a cavity or between studs without creating gaps or voids is, for all practical purposes, impossible to do. We need to design buildings that can be built easily. Similarly, we can't blame users for high energy use because they fail to modulate window opening in response to indoor air quality and wind speed; that is unrealistic and again good ventilation is primarily a design responsibility.

Now that there are thousands of buildings that actually perform as expected we can no longer blame the occupants for any performance gap. Some of these occupants may be self-build enthusiasts but others are care home residents or council house tenants with no previous experience of low-energy living.

It is not just about energy

Unsurprisingly, given what we have seen about energy, we find similar gaps between performance in other areas of building function. The analysis of the Grenfell Tower tragedy made it clear that the problem of substandard fire safety in buildings is widespread. Studies of indoor air quality in homes and schools have shown that current regulations do not guarantee a healthy environment. At best, this might mean some annoying mould issues; at worst, it means chronic medical conditions and death.

TOP Timber-frame structures are often designed using rules of thumb rather than calculating actual loads and this increases the amount of timber, double studs around openings for example, so that the timber fraction is often closer to 40% rather than the 15% often assumed. This reduces energy performance and increases construction costs.

ABOVE This is a well-constructed example of a partial fill cavity but even the small gaps will result in significant thermal bypass.

Form factor, massing and shape

In his book *Clean and Lean Management*, Dr Joseph Romm[1] estimates that around 85% of a building's life cycle cost is committed when only 7% of the project's costs have been spent. The exact figures will vary, but this highlights one of the fundamental problems with how buildings are planned and procured today.

The concept to planning stage is where clients and planners need to be wowed or appeased, but there is rarely the need to provide hard evidence of cost or performance beyond the essentials of access, parking and local planning constraints. It is a brave team that pitches a slightly improved version of a building that has already been proven in a similar or even identical form. A better business strategy is to promise everything in order to win the job and hope that extra budget can be found or corners get cut and *aspirational* performance targets dropped once the design process proper is underway. Any shortcoming between what was promised and what was delivered can usually be blamed on so-called 'value engineering' or the contractor profiteering.

Any budget estimate is likely to relate to an assumed cost per square metre of floor area. There may be little or no allowance for the additional cost incurred by complex shapes and inefficiencies of form, even though these are fixed at the early design stage.

While we cannot offer a formula to win an architectural competition with a proven and costed design, we can take a dispassionate look at how early design decisions can help deliver buildings that work and can be built on time and on budget.

Form factor

This simple ratio of external building envelope to useful floor area sets the energy use, cost and upfront carbon for a building for a given build quality. It tells us how much building fabric we need to build, pay for and heat or cool per unit of useful floor area. The smaller the number, the less we are paying in terms of upfront cost, carbon or land take to deliver useful accommodation.

The external envelope is where most of the cost is incurred. It is also where the heat is lost in winter, so this form factor becomes particularly important when we are trying to reduce heating demand or to meet a heating energy target that is based on the energy per square metre of useful floor area. Double-height spaces can be nice, but by removing useful floor area without reducing the heating demand of the building, we increase the heat demand per square metre, which is the target we are aiming for. We still have to heat the missing virtual floor area and pay for the building that encloses it.

The maths dictates that a double-height space costs more than double that of a single-height space per square metre of useful floor area. Of course, Passivhaus does not need to be a box, but if we want to deliver genuinely low-carbon buildings we need to think *inside the box* and stop apologising about houses that look like houses.

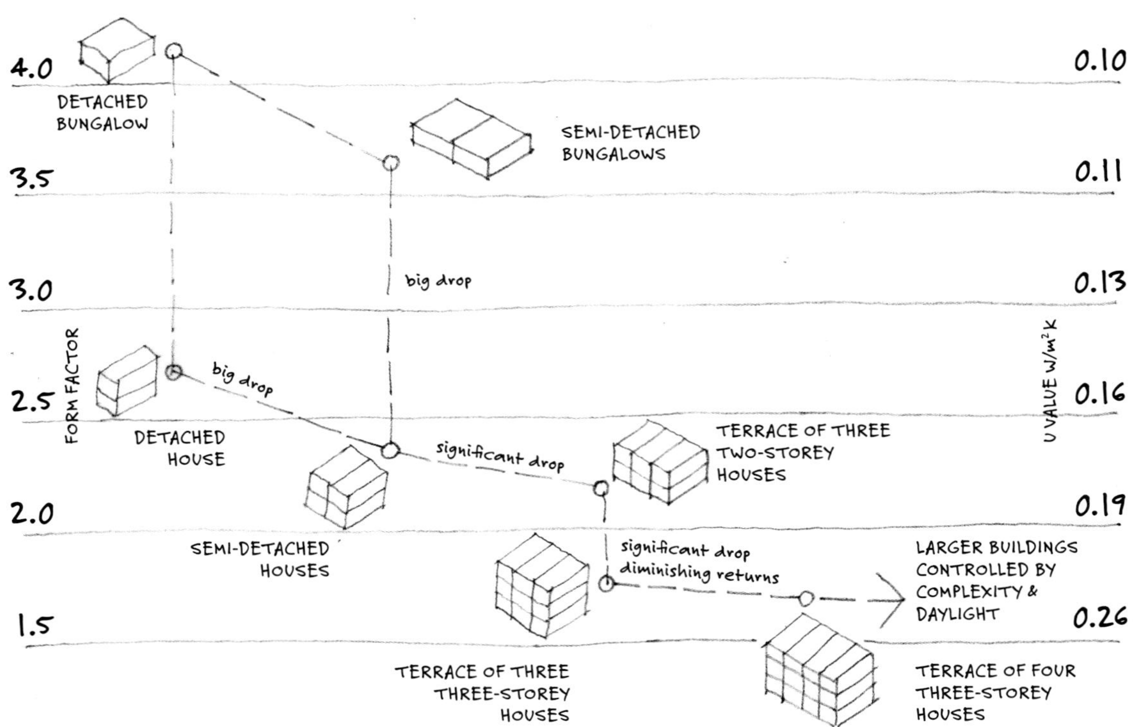

So, for rural and suburban houses we find that the two-storey semi is pretty good and still allows access to the rear of each property from the front. A three-storey building is even better but the cost benefits may be slightly offset by the fire protection requirements of the third storey. A terrace of three or more also improves the form factor but with diminishing returns and we need to provide either a separate access to the rear gardens or access through the house.

Historically, the driver for good form factor was cost and land availability, as we get more building for a given cost with semi-detached and terraced homes. The renewed interest in these traditional formats has been driven by the realisation that a lower form factor makes it much easier to achieve buildings with very low heating requirements and greatly reduced upfront carbon emissions.

Form factor depends on size and shape. A sphere has the lowest surface area for a given volume and this is the reason for the, albeit short-lived, popularity of domes for buildings. While the dome may be the simplest building form, it turns out to be the most complex in terms of keeping the rain out, fitting windows and doors or sorting internal layout and fittings – forget shopping for shelving at IKEA.

For larger buildings – whether schools, hospitals, office blocks or apartment buildings – the inherently bigger size compared to a detached house pretty much guarantees a good form factor. Indeed, as we saw, for anything much bigger than a three-storey semi, we are quickly into diminishing returns with respect to form factor.

For larger buildings, the heat-loss form factor ceases to be a big issue. Indeed, we want to stop the form factor getting too low as that means deep-plan spaces with not enough external wall to fit windows for views, daylight and ventilation. However, complexity still matters just as much. While form factor is easy to quantify, complexity is a lot more challenging. One architect's 'simple intersecting volumes' is an engineer's structurally and thermally complex 'interesting and expensive problem'.

Empirical guide to building design

The engineer and Passivhaus consultant Sally Godber came up with an empirical guide based on years of experience designing and certifying Passivhaus buildings in a wide range of materials and formats.

Sally has ranked various junctions from one, for a standard junction such as wall to floor, to ten, for a tricky detail such as an overhanging heated space. By multiplying the lengths of a junction by the difficulty factor and adding all these together, design alternatives can be compared. This is not a precise calculation but it is a useful tool to focus discussion at an early design stage.

More difficult junctions are inherently harder to build and tend to require more expensive and carbon-intensive materials and components. The cost difference is amplified if we care about heat loss and weather tightness. In situ, reinforced concrete freed up architects to design complex buildings with cantilevers and flat roofs. Direct glazed steel frames are an iconic symbol of New Brutalism. These were easy to build as long as we didn't need to keep the heat in or the rain out.

It may be that future AI-driven 3D printers using bio-based materials will allow new freedoms of expression without the issues of impossible-to-detail junctions. Such a technology will still be constrained by the laws of physics so a simpler building will still use less energy, be cheaper to build and provide more useful space for a given footprint.

Hardest to easiest	Relative difficulty*
Overhang	10
Balcony: inset	9
Inset doorway	8
Balcony: cantilevered	5
Wall-roof junction, parapet or gable	4
Deck access	4
Wall to roof below	3
Balcony: externally supported	3
Wall-floor junction	2
Wall-roof junction, eaves	1
Wall-wall corner (in plan)	1

Multiply by length of each junction to get score of overall complexity. Multiply by 5 if heavyweight façade.

RIGHT Sally Godber's difficulty rating for junctions assuming a good standard of thermal performance and minimal defects.

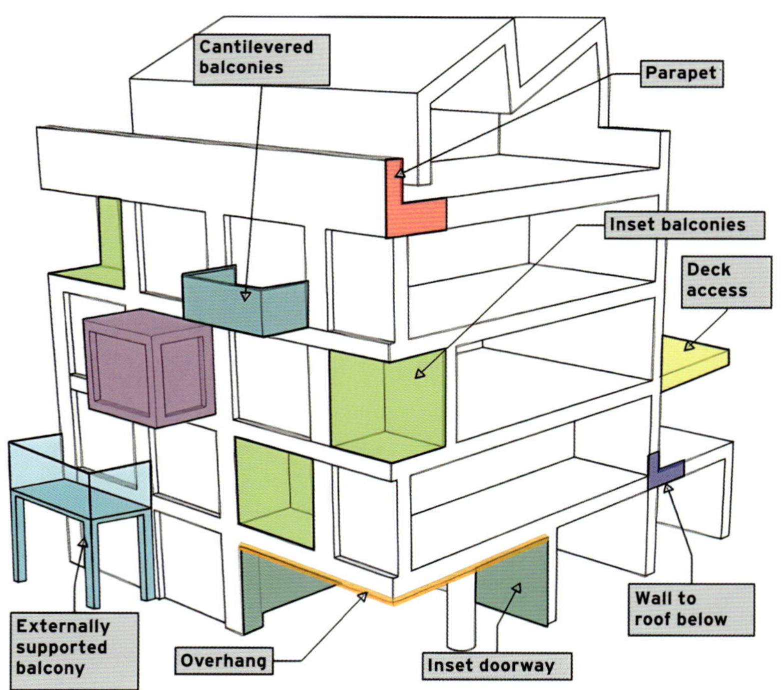

ARCHITECTURAL MINIMALISM VS ENGINEERING MINIMALISM

--

Art can bring comfort and relaxation but it can also be challenging, uncomfortable and even frightening. While we may seek out such experiences at the cinema, gallery or fairground, these are not generally qualities we want to welcome into our everyday lives. It's like the difference between a camping holiday and having to live in a tent because you don't have a home – it is rarely a lifestyle choice. While there can be fun to be had in making anything more complex than it needs to be, practical architecture and engineering often both aspire to some form of minimalism. However, architectural minimalism is generally the opposite of engineering minimalism.

Minimalist architecture tends to look very simple, despite being very difficult and expensive to build – it also requires great discipline and compromises to live with. A classic example might be the Edith Farnsworth House by Ludwig Mies van der Rohe. Visually, this glass-walled box of a house couldn't be simpler. But this simplicity is superficial and hides a lot of complexity behind every surface. It could be seen as a historic anomaly if such buildings were not still revered by designers today. Here are some examples of superficial simplicity that increase complexity and cost and reduce function:

- hidden gutters;
- concealed rainwater downpipes;
- cantilevers;
- parapets;
- anything with no visible means of support;
- shadow gap replacing skirting;
- avoidance of trim or architraves;
- corner windows;
- windows precisely aligned with ceiling or internal walls;
- fully glazed façades;
- pocket doors.

By contrast, minimalism in engineering or product design is very much about deep simplicity as a way to do more for less. Visual simplicity is not usually the aim – indeed, the result might look complicated. In nature we find the most complicated things in the known universe, yet most of these have come about by a very simple process of mutation and blind selection – albeit over an unimaginably long time.

ABOVE Farnsworth House, Illinois, by Mies van der Rohe 1951. One of many examples of superficial visual minimalism that hides extreme complexity under the surface. The authors have been guilty of this.

Environmental modelling and targets

For some time, low-energy buildings have been defined by vague measures, such as having thicker insulation, eco-materials or passive solar design, without verifiable targets to allow us to evaluate these claims. Quantifiable performance targets are crucial for driving improvements in any field. There are many performance targets that we can apply to buildings at the design stage, from fire resistance to ventilation and acoustic performance. Here we will just focus on energy use and operational carbon emissions to explore some of the issues raised.

If we look at energy use, whether measured or predicted, then we can compare with benchmarks or set targets. To do this we need to convert the energy use for a given building into a form that allows comparison between different buildings. A number of buildings standards divide the annual energy use by the floor area of the building to give a $kWh/(m^2a)$ figure. The Passivhaus standard uses this approach, so a home and school might both have an annual heating demand of around $15kWh/(m^2a)$. While the headline figure might be the same, the split between the various losses and gains will be slightly different for a school or house. However, useful comparisons are still possible due to the simple trick of converting all heat gains and losses to a specific per square metre figure. What is more, we can be confident predicting the heating and cooling energy use or loads for a completed building of a given size, even before we have started to design it. Being confident that the peak heating demand for a $3,000m^2$ school will be close to $30kW$, even before engineers have been appointed, allows early decisions to be made about space required for heating plant and radiator size and location.

Where this energy per unit floor area approach gets criticised is that a house the size of a small school will be shown to be as efficient as a small house built to the same standard. But the energy use per occupant – not to mention the upfront carbon emission – will be much higher.

An arguably more appropriate metric is energy use per occupant. This is useful when journalists write about the environmental performance of a building but it is problematic for regulatory purposes for at least two reasons. How do we know how many people a house is designed for? Do we count bedrooms and assume an occupancy? All the evidence suggests that smaller homes have higher occupancy than larger ones. Is the room labelled as a fifth bedroom evidence of high occupancy, or where the wine fridge will be located?

We could use a standard occupancy algorithm as used in many energy calculation tools to estimate occupancy and thus work out internal heat gains. This would reward smaller buildings as they will be shown to use less energy per person, as we might expect. However, if such a metric is to set targets for regulatory compliance we would have to make judgements about how large a dwelling is reasonable when setting the compliance target.

The problem is not the metric, which is what we are considering here. Larger buildings will use less energy per square metre of floor area just as larger fridges of the same specification will use less electricity per unit of storage volume. A 100mpg scooter with one rider is less efficient than a gas-guzzling 40mpg people carrier with an average of three passengers. It is the same for buildings.

We can't look to energy models to address issues of equality or fair shares of carbon budgets. What we can do as designers, clients or critics is call out oversize projects that claim to meet any environmental standards. If we care about keeping the planet habitable, no amount of efficiency can compensate for too much stuff. It might be convenient to draw an imaginary boundary around our PV and lithium battery-powered, mass timber, negative carbon building, but as we broaden the boundary to include the whole planet, we realise it is carbon all the way down. No cradle-to-cradle carbon accounting can look at the whole picture because the whole picture is too big to model or even contemplate. While the almost infinite web of supply chain connections may be impossible to model, half as much stuff will have about half as much impact and will cost about half as much. This may seem overly simplistic but it is a useful reality check if you find yourself disappearing down the rabbit hole of creative carbon accounting. We look more at this in Chapter 3.2.

A metric with unintended consequences

Another method is to set a percentage improvement target. At the time of writing, this is still the basis for meeting the heating energy requirements of UK building regulations and standards. A number of environmental auditing tools use this approach to encourage year on year improvement. The now abandoned Code for Sustainable Homes (CSH) in the UK used this approach with code levels 1–6 based on a percentage improvement over the same 'notional' building of the same size and shape designed to a base specification. While a floor area metric is seen by some to encourage oversize buildings, the percentage improvement approach has a more fundamental flaw in that it favours inefficient forms.

By way of illustration, one social housing project at design stage was just achieving code level 4 as required for funding. The energy consultant, a practical self-builder, suggested a change to the floor plan that simplified the structure, reduced heat loss, saved cost and provided more useful floor area for the residents (see Chapter 3.2, which looks at the fundamental impact of form factor on a building's upfront and operational energy requirement in any climate). This should have been good but because the CSH used a percentage improvement target rather than an energy use per unit floor area metric, we were now failing to meet the required target, even though actual energy use would have reduced. The solution provided by the code assessor was simple: switch to direct electric heating as that would increase the emissions and running costs so that the percentage improvement required could be more easily achieved. This would have meant that the emissions and running costs would be higher than our non-compliant version. Similarly, it was quite easy to achieve the percentage improvement needed for level 4 for a large detached property but almost impossible for a compact apartment.

Models and tools

When Nick designed his own home, he didn't do any energy calculations other than U-values, but this was 1995 – some four years after the first Passivhaus in Germany and 18 years after Harold Orr's Saskatoon Conservation House (1977) – so we were aware of the main ingredients needed to make a low-energy home.

The walls were straw bales, the roof had 400mm of blown recycled paper between thermally broken homemade I-beams and the floor floated on a continuous 100mm-thick layer of expanded polystyrene that was continuous with the wall insulation. What is more, airtightness and thermal bridges had been thought about in some detail. The windows were some of the best-performing double glazing available. Frames suitable for high-performance triple glazing were not easily available in the UK at the time. We installed the windows in line with the timber cladding to make sure any leaks around the windows or sills wouldn't get into the straw (unaware that this greatly decreased the thermal performance of the installed window).

So, why bother modelling it when all the ingredients were about as good as we could get? In hindsight, we had good ingredients but we were not following a proven recipe. Most people know that basic cake can be made from eggs, flour, butter and sugar. At first glance there doesn't seem to be much to baking, surely not enough to glue a nation to the TV watching people combine these ingredients each week. As when baking a cake, the proportions of ingredients are crucial and creative choices; a squeeze of lemon juice, a dash of vanilla or a beautiful window can elevate your creation.

As we saw earlier, when making a cost-effective building the key proportion that matters is the ratio of external building fabric (walls, ground floor and roof) to the useful floor area of the building: the form factor. For our supposedly low-energy house, the ratio was over five – which it turns out, is not good.

RIGHT With thick insulation, good airtightness, natural ventilation, solar hot water and a zero-carbon woodstove for all heating, Nick didn't see the need to model his self-build home! How wrong he was.

Rules of thumb and doing some simple numbers

A reasonable target for the form factor of a detached home is around three (this assumes a heat-loss area measured externally divided by the treated floor area measured in accordance with Passivhaus methodology). Other area conventions will give a slightly different number that won't be directly comparable. If we are aiming for Passivhaus levels of performance, as a rule of thumb for our UK climate, the wall insulation thickness in millimetres needs to be around 100 times the form factor. This assumes basic air-based insulation with a conductivity around 0.04W/mK. Our straw insulation was 360mm thick with a conductivity around 0.05W/mK so we would need a straw insulation thickness around 5 x 100 x 0.05/0.04 = 625mm - nearly double the thickness of our bales on edge. This thickness also increases the external heat loss area or reduces the useful floor area, making the form factor worse!

Unfortunately, Nick is doing these simple calculations nearly 30 years after designing and building his home. We can upgrade our windows, ventilation and heating system, but we are stuck with the high form factor. Interestingly it was our use of the pattern language that led to the inefficient shape. As the pattern language is an empirical tool, we can update it to include form factor, but we can also check out the impact of any design option with some simple calculations or a more detailed model.

Glazing

After the size and form factor, glazing is almost always the next most important consideration for most buildings. Again, there are useful rules of thumb that balance daylight and summer overheating risk. Such guidelines, calibrated from experience of repeating the same pattern over thousands of buildings, have served well for much traditional architecture. See the next chapter for more discussion on glazing.

LEFT Jon designed his home following rules of thumb in the early 1990s but finds it would not be approved under current regulations.

Simple models

Created in 1931 by Harry Beck, the London Tube map is a design icon with proven functionality. As a map it is wrong and it would be very misleading if you wanted to get around London on foot! However, if it were more accurate as a map it would be unusable as a way to navigate the complex Tube and rail network. As the saying goes, 'all models are wrong, some are useful', and sometimes they have to be wrong in order to be useful. It is the same with calculating building structure or heat loss; it is more useful if we focus our efforts on what matters and do not get too lost in details that do not. But knowing what matters is not always obvious without some calculation or monitoring.

The form factor rule of thumb for wall insulation thickness is useful because it is so easy and encompasses many considerations, but it is a simplification. For example, while the walls usually represent most of the heat loss area of a building there are also roof and ground as well as ventilation losses. Then there are internal and solar gains, which will vary with building type and climate. Also, there are specific wall insulation thicknesses that are easier to construct, so we would not want to use the rule of thumb to work out actual insulation thickness, even with adjustments for conductivity as we did for the straw.

For example, some Passivhaus flats in London had a form factor of around two, suggesting 200mm of insulation. The chosen construction was a masonry cavity with brick skin. That was easier to build if the full-fill cavity could be limited to 150mm so that set the insulation thickness. The insulation could have been changed to one with a lower conductivity to give an equivalent 200mm at $0.04W/m^2K$. But cost and other issues, such as thermal bypass and fire risk of lower conductivity rigid board insulation, need to be considered. The point is that the simple model was a really good starting point for design and also provided reassurance for those expecting the 300mm of insulation that was previously assumed to be required for a Passivhaus in the UK climate. If having optimised the form we can also optimise the glazing, based on simple daylight calculations or rules of thumb, then we should not be far off having a building that performs very well. This is crucial if, as is often the case, planning is obtained before more detailed energy modelling has been carried out.

However, we may want to push the boundaries, forcing us to abandon our rules of thumb for any number of reasons, from a design whim to fixed constraints of the site or planning requirements. If this is the case, we will need more sophisticated models and our building will almost certainly be more expensive to build if we are to maintain the same performance and function.

Dynamic modelling

One way to model the energy use of a building is to use hourly historic weather data. Suddenly, our simple sums become very involved. This approach is called dynamic modelling and it can be useful for answering certain questions or for testing simplified calculation methods. But like the Tube map, simpler models have far higher utility and are less likely to be wrong, or are at least much easier to check. As philosopher Carveth Read

once said in a quote commonly misattributed to John Maynard Keynes, 'It is better to be vaguely right than exactly wrong'.

The good news is that we get remarkably good results if we just use the average monthly external temperature and assume an average fixed indoor temperature. The daily fluctuations and slow seasonal changes mean that the heating and cooling of the building materials hour by hour cancels out. This means we just need to do 12 simple monthly calculations to model the heating or cooling demand of our building. This is the basis of the calculations used in most national building codes such as SAP in the UK. A spreadsheet usually takes care of the 12 calculations, which all use the same building inputs as the areas, and U-values remain the same.

In fact, if we just use a single calculation based on an average heating season temperature difference and time, we get a pretty accurate result. This is useful for doing a quick reality check. In theory, the hour-by-hour calculation should be a better reflection of reality but, as with a topographically correct Tube map, that precision comes at a high price.

The Passivhaus Planning Package

The building energy model of choice for a growing number of building designers and energy consultants is the Passivhaus Planning Package (PHPP). This is a monthly energy model (a large spreadsheet) with climate data for much of the world already built in. More information is available on Passipedia and Sarah Lewis's excellent book, *PHPP Illustrated*.[1]

The PHPP evolved as a tool for designing Passivhaus buildings, which aimed to solve the problem of low-energy buildings performing much worse than predicted. Part of the solution was the realisation of the importance of thermal bypass and thermal bridges, as well as build quality. However, it was also clear that assumptions used in modelling would have a big impact on our design decisions, such as how much insulation we need or whether heat recovery is worthwhile.

2.5

Passive solar design

Along with form, fenestration is probably the most important consideration in terms of what delivers good thermal performance. When considering orientation, facing a building towards the equator (south in the northern hemisphere) would seem like the obvious thing to do. This is the basis of solar architecture, which has been around for thousands of years. Over 2,000 years ago, the ancient Greek philosopher Aeschylus wrote, 'Only primitives and barbarians lack knowledge of houses turned to face the winter sun'.

More recently, we have seen the advent of modern solar architecture using large areas of glazing and storage of heat in massive concrete structures, huge tanks of water, or basements filled with rocks with heat transferred by big fans and ducts (see Chapter 3.1 for more on the thermal mass debate). Many examples were built in the 1970s and 80s. Whatever enthusiasts may claim, they did not work – they were hot in summer, cold in winter, and expensive to build and maintain. There were other problems such as very constrained form and the need for suitable sites with good solar access.

What did work was superinsulation. Also in the 1970s, largely unsung pioneers in Scandinavia and North America found that by building carefully insulated, airtight buildings, the heating could be mostly provided by the heat given off by people and appliances. With such low heating requirements, it made no economic or environmental sense to try and meet that small amount of demand with extra glazing and thermal storage. While the heating demand was low, it was occurring at the coldest time of year with the least solar gain, so collectors would have needed to be very oversized to grab those weak winter rays.

In his article 'Solar Versus Superinsulation',[1] Martin Holladay makes it clear that the technical debate between superinsulation and passive solar design was effectively decided by the end of the 1970s. However, we still see old solar architecture ideas reinvented as 'an architectural response to climate change'. Solar architecture seems so much more interesting and 'architectural' than good insulation, airtightness, heat recovery ventilation or the avoidance of thermal bridges. As the urgency of the 1970s oil crisis appeared to wane, radically reducing building energy use moved from urgent necessity to aspiration and lifestyle choice.

It took the heightened awareness of the climate catastrophe and the advent of Passivhaus in the 1990s to rediscover and refine the superinsulation approach and provide a focus for continual development.

Despite the somewhat misleading name, Passivhaus came from the school of superinsulation rather than passive solar. Where we can orientate our building to face south, low winter sun will animate any interior with sunlight and some gentle warmth. A roof overhang and deep window reveals will also provide shading from summer heat. In northern climes where the midwinter sun only just clears the horizon, a southern aspect lets the sunlight in to lift our spirits, even if it doesn't do very much in the way of heating. If we are lucky enough to have a south-facing (in the northern hemisphere, or vice versa) site it frees up our design options should we have the budget for some generous glazing. Even without additional shading overhangs, south-facing glazing will naturally limit summer gains simply because of the angle of the sun.

Hope View House

Award-winning Hope View House by Warren Benbow Architects is a nice example of an equator-facing Passivhaus home that is warm in winter and not too hot in summer, thanks to the generous overhang and due south orientation. However, even with fixed shading, the diffuse solar gains are significant and some effort is required to maintain good summer comfort using night purging. There is also an increased risk of summer discomfort due to radiant temperature asymmetry, as explored in Chapter 2.1.

Heating energy and power - gains and losses

The following graphs are derived from the PHPP design tool (a spreadsheet). The first shows the energy balance for heating over the year, with losses in the left column and solar and internal gains on the right. The heating energy makes up the balance, as gains must equal losses unless the house is cooling down or heating up. The second graph is the balance of heating power on the coldest day of the year.

As the gains and losses are all shown per square metre of useful floor area (called treated floor area, or TFA - clearly defined in Passivhaus methodology), we can compare the various gains and losses for very different buildings - from a small house to a large school, for example.

With experience, we can tell a lot about a building design just from these graphs. In this example the glazing dominates the losses and provides a significant proportion of the (very small) heating demand over the year. The peak heating load graph shows that, while during the coldest days of the year the windows are the main heat loss, they do very little to keep the building warm at this time - hence we still need a heating system.

Passive solar heat - not such a free lunch

From the first graph we see that solar gain is providing nearly twice as much heat as the heating system. But let's look at the numbers. In this example, the annual solar gains

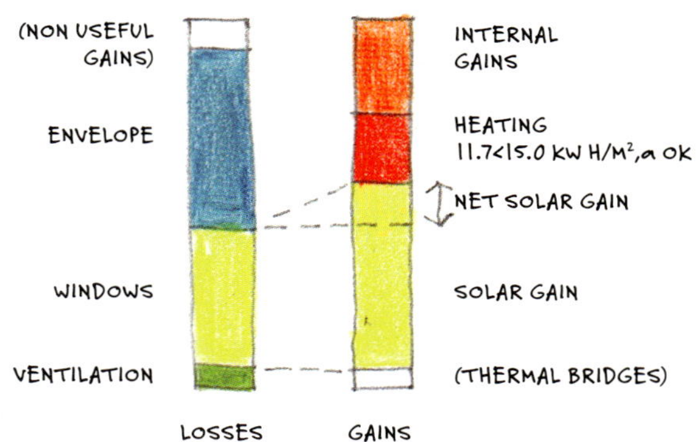

ANNUAL ENERGY BALANCE
KILOWATT HOURS PER SQUARE METRE PER ANNUM KWH/M²,a

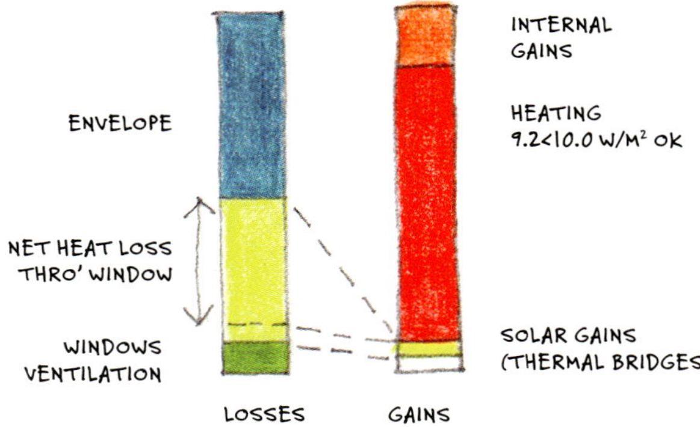

HEAT LOAD BALANCE ON COLDEST DAY
WATTS PER SQUARE METRE W/M²

SOUTH FACING HOUSE

through the south glazing are calculated as 4,779kWh/a. With transmission losses of 3,182kWh/a, we have a net gain of 1,597kWh/a from 52m² of high-performance triple-glazed sliding doors. The heating is provided by a heat pump with a coefficient of performance (COP) of over three so at 20 pence/kWh electricity cost (energy prices are unstable but construction materials such as windows have also increased significantly), we are saving about £110 a year, thanks to passive solar gain. That is about £2 a year for every square metre of south glazing, which will cost upward of £500 plus installation. If we add in shading from external blinds or overhangs then that can easily double this capital cost while adding more upfront carbon emissions to harvest a very small amount of useful heat.

We need some of that glass anyway for daylight and views, so the useful heat gains are, in effect, free. Indeed, even east and west glazing can provide a net heat gain. However, if we are adding glass as a way to heat our houses, it is probably the most expensive heating source available. But notice we still need heating (the red part of the graph). Adding more glass does very little to reduce this as the match between heating demand and solar gain is completely out of step. Instead, as we add glass to try and grab those weak winter rays, we lose more heat on the coldest days and it becomes more and more difficult to stay cool in summer. Sure, we could add external blinds, insulated shutters and some form of inter-seasonal heat store with computer control. Before getting too seduced by the potential for innovation, remember that the heating bill for this large 260m² detached Passivhaus averaged at about £4 a week over the year in 2020.

Although this may be disappointing news to solar architecture enthusiasts, others will be excited to realise that in order to build an energy-efficient building we don't need to limit ourselves to the small number of south-facing rural sites with good solar access.

Once we move beyond single family dwellings to look to non-domestic buildings and multifamily homes, the potential to face all habitable rooms south becomes impractical and so we need to make all orientations work. Achieving a low heating demand becomes easier because the free heat gains increase and heat losses reduce, thanks to the better form factor.

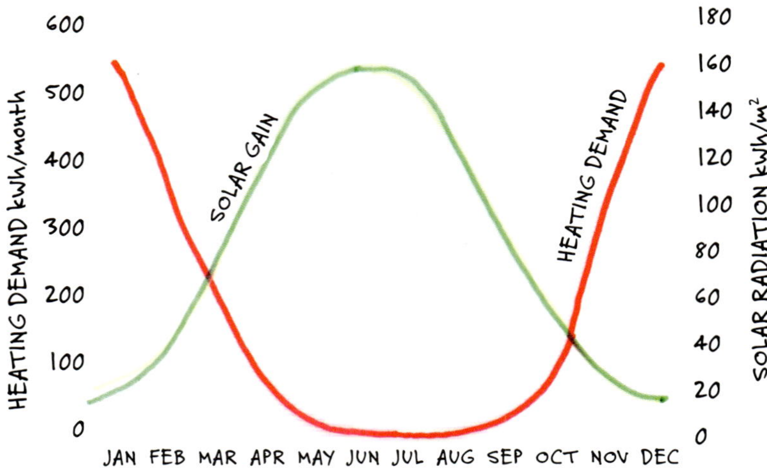

RIGHT This graph shows the extreme mismatch between heating demand and solar gain.

Challenging plot is a lightbulb moment

A self-build client had spent a considerable time searching for a south-facing site close to a town centre to build their dream Passivhaus home. A serviced plot was eventually found on a new estate within a few minutes' walking distance of Bishop's Castle town centre. Unfortunately, the houses and roads had all been laid out with neighbours to the north and south, leaving views to the west and the street to the east – far from the passive solar ideal. Modelling in PHPP confirmed that we would have to keep the east and west glazing modest to ensure good summer comfort, but also that we would struggle to hit the Passivhaus heating demand target of 15kWh/(m²a), so easily met by the south-facing example above.

The east–west orientation meant we had to work harder to maximise daylight and views while using the minimum of glazing to avoid overheating. This had the benefit of reducing heat loss and cost because even expensive triple-glazed windows lose about six times more heat than the walls.

Floor-to-ceiling glazed doors gave way to a square window with window seat, and access to the garden was through a glazed door on the north that provides modest infill daylight. The kitchen sink and window were to the east, providing connection to passers-by when standing at the sink but minimal overlooking due to the high sill height. Splayed window reveals maximised the daylight quantity and quality for a given window size, as did light paint colours for the walls and ceilings.

The site constraint could have been seen as a disappointing compromise but instead the result was a pleasant surprise for the owners who really enjoy the morning and evening sun and the window place that frames the view.

LEFT Bishop's Castle Passivhaus by Dempsy Decourcy Architects. What started as a challenging west-facing plot led us to fully embrace the idea of designing for daylight and views, not winter solar gain.

Design for peak heating load

Although the final daylight-optimised design misses the Passivhaus annual heating demand target by about 2kWh/m²a, it does meet the so-called 'alternative criteria' based on a maximum peak heating load of 10W/m². Although the definition of a Passivhaus building is based on the peak load, there has been a tendency to see this as a second-best solution only for difficult sites such as this. Similar heating load targets predate Passivhaus and we now think the Passivhaus peak load target is more appropriate generally.

As we have seen, extra glazing only gives a small reduction in the cost of heating but increases the peak heating load on the coldest days. This means that if we design for a heating load target, excessive glazing is penalised rather than encouraged. By comparison, it is common to keep increasing south or even west glazing to nudge a design over the line when using a heating energy target based on kWh/(m²a).

We can compare the following graphs for the west-facing Passivhaus home with that of the highly glazed south-facing example of Hope View House. Experience shows that where the gains in the energy balance graph are not dominated by solar gain, the building will be more comfortable in summer. Generally speaking, though, there is less heat gain from windows. There is also less heat loss due to glazing, so the energy equation remains in balance without large south-facing windows. The same is also true when considered from a heating load perspective.

In terms of carbon emissions, for a fossil-fuel-heated building energy use is more relevant than peak load. The power of the heating system is less important than the amount of gas burnt. As we move to 100% renewables, it is the capacity to meet peak loads and to match demand and supply that becomes the more important challenge. The wind and sun may be free, and the cost of photovoltaic panels and large-scale wind turbines falling, however, being able to deliver renewable heat on demand when the sun is not shining and the wind is not blowing is expensive.

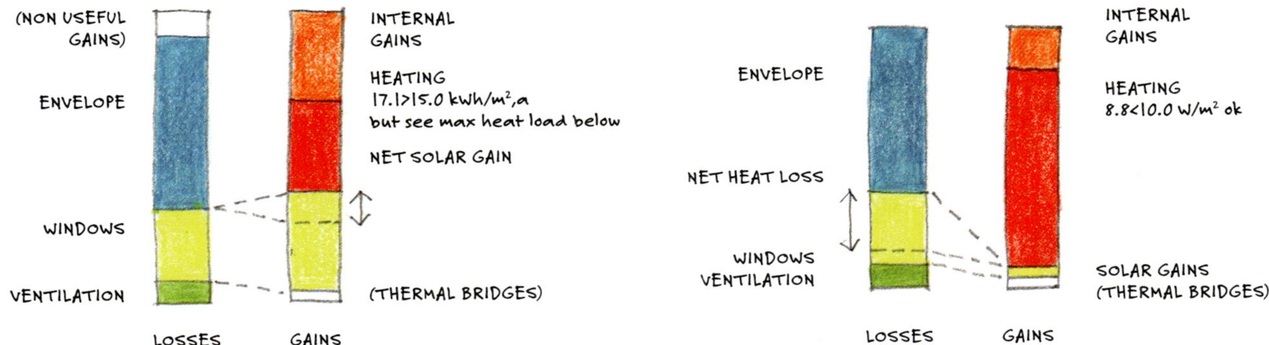

The 10W/m² peak heating load criteria of Passivhaus was based on the idea of reducing the heating load to a level that could be met by heating via the fresh air required for ventilation but without recirculation. The theory was that this would eliminate the need for a conventional heating system and so save enough capital cost to pay for the extra insulation, heat recovery ventilation and triple-pane windows.

We do not recommend heating with fresh air but are happy that the experience from thousands of buildings shows that this possibly arbitrary number of 10W/m² is still a pretty good target to design to. The idea of such a target is that it is largely independent of climate or current grid carbon values or energy costs.

Beyond our professional experience, there will be climates and situations where solar architecture works: where temperatures are not too extreme, the sun mostly shines and land is cheap.

Heading north to Scotland, the winter days are shorter and the summer days longer, so passive solar makes even less sense than in southern England. Again, what does work is insulation and fenestration designed for daylight and solar control on the long days of surprisingly strong summer sun. Overheating buildings in Scotland become a reality once we have a bit of insulation and more generous glazing. The problem is made worse by clients and designers living in older, uninsulated homes with smaller windows, who cannot imagine overheating being a concern until they experience it first-hand.

Interestingly if we were to move our west-facing house in Shropshire to Dundee, the same building still meets the peak heating load criteria for Passivhaus, suggesting the design is likely to be robust and relatively insensitive to the uncertainties of weather and climate change. While the west-facing example above required some effort to optimise the daylight without incurring overheating, designing for daylight over winter solar gain is a good approach for any building. Hopefully, you are now convinced that designing for passive solar heating is an outdated and flawed concept for all but a few niche situations.

BELOW There may be places where passive solar homes work?

Radiant asymmetry

Avoiding discomfort from cold radiant temperatures in winter is a key principle of Passivhaus and is the main driver for requiring triple-glazed windows in the UK climate, even if the energy balance would allow double glazing. However, summer overheating assessments just consider air temperature, specifically hours above a limit such as 25°C in the case of Passivhaus compliance.

When Nick was working on a school with full-height glazing in the classrooms, some people reported the 'hot, stuffy and airless' conditions. The glass was almost fully shaded in summer by fixed overhangs but there was sufficient solar gain to warm up the glazing to around 40°C. Convection and conduction ensured that the whole glazed area warmed up. The shading was reducing the gains to the room and the air temperature was around 23°C, which would normally feel quite comfortable on a summer day. CO_2 readings suggested good ventilation but the perception was still 'hot and stuffy'.

More recently, a client reported the same 'hot, stuffy' experience with a detached dwelling. The client had pushed for larger glazing areas than we recommended but the PHPP, with conservative assumptions, indicated good comfort in terms of overheating hours. On visiting the home in June, the air temperature was also 23°C with ventilation by mechanical ventilation with heat recovery (MVHR) and open windows, but the dining area in the kitchen was undeniably uncomfortable and, like the classroom, felt 'stuffy', despite plenty of fresh air.

The thermal image shows the surface temperature of the east and south glazing. Again, shading was reducing most of the gains but there was sufficient direct, diffuse and

reflected radiation to warm up the glass enough to cause radiant discomfort. Unlike the classroom situation, it was very easy to demonstrate the effect by simply walking to the other end of the room where the air felt fresh but had the same air temperature, relative humidity and CO_2 levels as the 'hot, stuffy end'.

Humans have evolved to be very sensitive to our thermal environment and what we feel as warm or cold depends on air temperature, air movement and the temperature of the surfaces around us, which changes how we lose or gain heat by radiation. Relative humidity also has an influence, particularly as we start to feel too warm.

Modelling the room with a simple online 2D program suggested that the mean radiant temperature for a person stood in this uncomfortable zone was not high enough to explain the discomfort due simply to high temperature. The discomfort seems likely caused by something called 'radiant asymmetry', rather than just too warm a temperature. Radiant asymmetry is the difference in temperatures experienced by two sides of our body due to radiant heat transfer with surfaces that are at different temperatures.

While easily experienced, modelling and even measuring radiant asymmetry and the impact on the comfort of an individual is actually quite challenging. Most research has focused on winter discomfort caused by cold surfaces, such as glazing that is fixed by the use of triple glazing or even curtains.

Until readily usable calculations or design guidance are available, we can at least be aware of the principle, which will allow us to avoid most situations where this is a problem. Without understanding what is causing the discomfort, we may try and solve the wrong problem, for example by cooling the air or increasing ventilation.

Window design from the inside out

Having looked at the implications of designing for solar gain, we will now look at other aspects of window design: daylight, sunlight, ventilation, view and the impact of windows on the interior and exterior of a building.

To obtain good daylight in a room up to 6–7m deep with a single window, size it to be around 20% of the internal wall area that the window is in. Note that this is often well in excess of the statutory minimum of 10% of the floor area, which is often remarked on by building users as creating gloomy conditions. Rooflights will provide more than twice as much light as a window and can light large rooms, but beware of excessive solar gain. Rooms with light from two directions reduces glare and leads to a much better visual environment and sunlight penetration at different times of the day – a feature recognised in Christopher Alexander's *A Pattern Language*[13] as 'light from two sides of every room' (Pattern 159). Creating plans that enable windows on two sides can lead to layouts with many offsets with resultant increases in the form factor (see Chapter 2.3) and cost.

Sunlight enlivens interiors and comes from different directions, but be aware of the risk of overheating. Shading from an overhang to south-facing windows can reduce sun penetration in summer, while still allowing the sun to penetrate in winter. The critical orientation in summer is often the west, which receives low-level sun in the late afternoon. It is also difficult to provide adequate shading against glare and excess heat.

BELOW Passive night cooling can work well but the reality is often different to the design aspiration and airflow rarely obeys the designer's arrows.

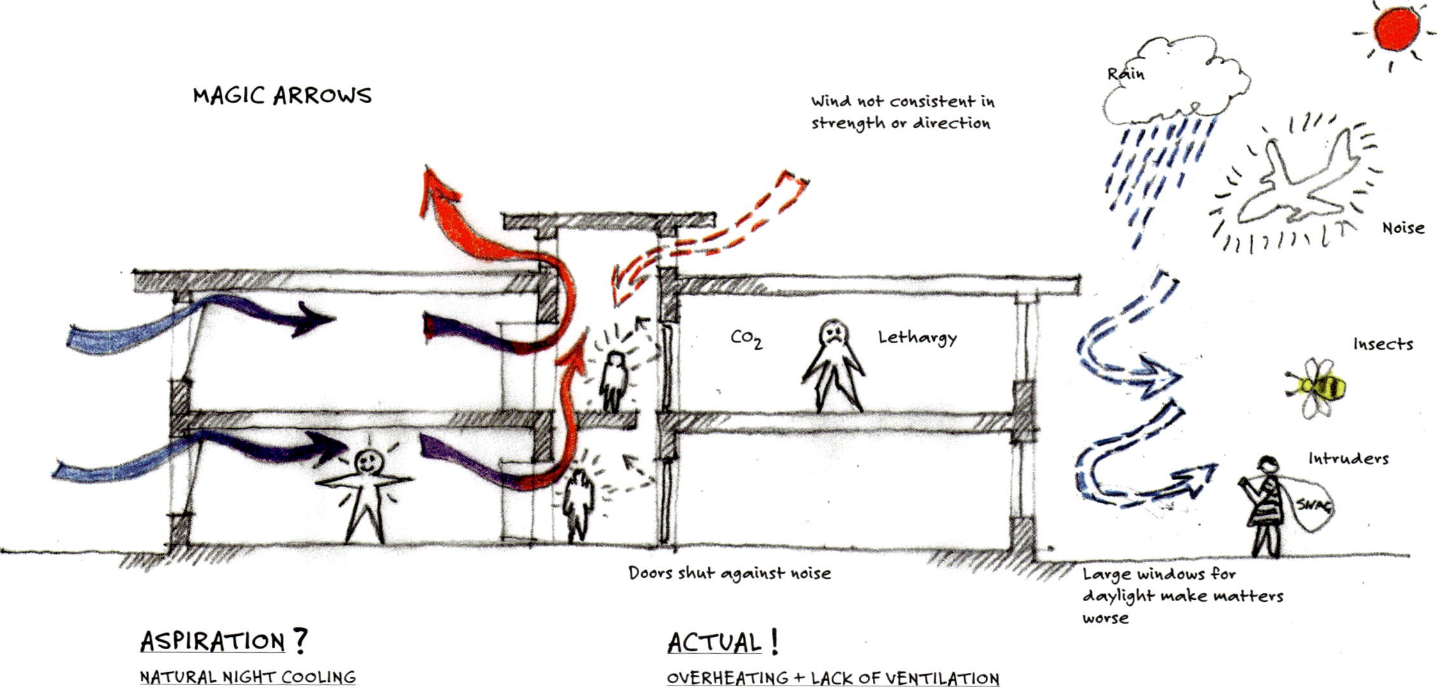

A single window provides natural ventilation (see Chapter 2.8), again to a room with a depth of 6–7m. The design of an inward-opening tilt and turn window provides secure, weatherproof night ventilation. Cross-ventilation is desirable and obtainable from opening windows on two sides of a room.

Jon has windows on two sides for the benefit of the room but regrets only making one openable to save money. His house also has double-height spaces (see Chapter 2.3) with very little ventilation at high level, which means that in summer stagnant hot air accumulates where high-level ventilation would promote the stack effect from hot air rising to draw cool air in at low level.

That said, it is always best to design out summer overheating by controlling gains as the ideal of night purging is not always possible. This may be because of noise, insects, security concerns, absence of occupants (summer holiday or unoccupied flats) or because in a proper heatwave, the air at night may not cool below 20°C.

There is a temptation to design a large picture window facing a view, though this will have implications on heat loss and overheating risk. Small windows focused on a distant hill or other landmark will make the enjoyment of the view a conscious act in its own right. Windows serve to make a connection with the outside world and this can be promoted by keeping sill heights low so that the foreground is included in the view; 400mm is coincidently around the right height for a window seat. The glass can also be a little higher than the seat with a solid panel under the sill. Bay windows can create a special window place.

Jon has first-floor windows with the head height below average eye level, at 1,450mm above floor level, and a low sill height. This directs the view down into the garden. Note that we do not advocate glazing down to floor level as this increases heat loss and overheating risk for little view, daylight or sunlight benefit. Consider privacy; in this connection a raised sill level can prevent people seeing in a window on a busy street. Remember that external doors are part of this discussion and that glazed doors are often cheaper than solid insulated ones. Glazed internal doors and internal windows afford connection between spaces that can make a small building appear much more spacious. Where privacy is needed, the traditional detail of a window above an internal door serves a similar purpose and can indicate if the bathroom is occupied without having to knock, or that you forgot to turn the light off.

ABOVE This window provides daylight and views of trees and birds but does not need to open as it would be difficult to reach. Other windows in the room do open allowing cross ventilation if required.

ABOVE 'The best window in the house'. Careful placement of the bedroom windows allow views out from the bed without people being able to see in.

RIGHT Thanks to the thickness of walls and floors, windows that look generous from inside can look a lot smaller from outside. The large window and door are just visible to the right of the external image.

OPPOSITE Slate House, a Passivhaus self-build in Herefordshire. Viewed from inside, the windows feel very generously sized and in hindsight the owners think the house is actually a bit over glazed.

Design from the outside in

Along with designing to maximise solar gain, another common and problematic design approach is to use fenestration to decorate buildings. As with free heat from windows placed for daylight, we will get interest and potentially even visual delight from fenestration placed to serve another function.

Problems occur when we size and place fenestration to create a composition from the outside. First, there is no reason why what works for an external visual composition should work for the occupants looking out - indeed, when it comes to privacy, the opposite is likely to be true. Second, thanks to the thickness of walls, floors and roofs, the internal wall area is generally a lot less than external. This is particularly true of smaller, well-insulated homes. In these cases, what may look like undersized windows from the outside may well feel generous or even oversized from the inside.

Good design principles suggest that windows should be placed to serve the building occupants. Some of these requirements are quite general, such as natural daylight, ventilation and escape. Others will be more site specific, such as views or the need to maintain privacy without drawing the curtains. Having looked at the design in detail from the point of view of the building users, you need to consider what the building looks like from the outside - this is where personal taste comes to bear.

Having discussed how windows fit into the design of the building as a whole, we consider the design of the windows themselves in Chapter 3.5.

Why Passivhaus?

Passivhaus is a building standard that aims to be truly energy efficient, comfortable and affordable at the same time. Dr. Wolfgang Feist, who co-developed the Passivhaus standard with Bo Adamson, explains how it came about:[1]

> *I was working as a physicist. I read that the construction industry had experimented with adding insulation to new buildings and that energy consumption had failed to reduce. This offended me – it was counter to the basic laws of physics. I knew that they must be doing something wrong. So I made it my mission to find out what, and to establish what was needed to do it right.*

In 2007, just as we were getting to understand the complexities of each new version of EcoHomes, the Code for Sustainable Homes (CSH) was launched in England. The serious flaws and anomalies in the CSH became obvious as we tried to apply it to energy-efficient designs we were already working on. Higher levels of the Code required on-site generation and we saw a short-lived growth in micro wind turbine manufacture and installation on urban rooftops.

The final straw was when a particularly well-designed housing development achieved Code level zero because it missed the required water efficiency target. The irony was that the buildings had genuinely state-of-the-art water efficiency measures. At the time, Nick and others wrote critiques of the flawed energy and water aspects of the CSH, while coming under fire from fellow greens who welcomed the CSH as a bold step in the right direction that we should not criticise openly. Nick's critique of the water and energy aspects of the code in *A Critique of the CSH Water Efficiency Requirements*[2] did lead to some changes, but the whole process was a huge effort and a lost opportunity to promote measures that would deliver real results.

ABOVE The range of 'green' standards can be overwhelming and many turned out to be seriously flawed. This is why we approached Passivhaus with considerable scepticism and caution.

RIGHT Sigma Home at the UK BRE Innovation Park. The first building to reach level 5 of the Code for Sustainable Homes 2007.

In 2007, Nick went to the International Passivhaus conference in Bregenz, Austria. Disillusioned by the UK situation, a number of architects and energy consultants were looking to the Passivhaus standard as a better way forward. The contrast between what was happening in the UK compared to what we saw first-hand in Austria and shared in conference presentations from further afield, was stark. What we saw was a considered and proven approach, informed by evidence and hard numbers rather than tokenistic visual adornments and eco-bling. Where technology such as heat recovery ventilation was used, it was designed and installed properly to maximise efficiency and comfort. Meanwhile, in the UK we saw noisy makeshift installations with flexible tumble drier exhaust hose and duct tape in supposedly exemplar CSH projects.

Writing this chapter nearly 20 years on, EcoHomes and the CSH are gone and there are now thousands of Passivhaus buildings in the UK. The Scottish Parliament has just announced a plan to require Passivhaus or equivalent for all new homes. While any number of unforeseen events could lead to such a bold policy being diluted by the time this book is in your hands, the fact remains: there are enough built and occupied examples out there to dispel all the early myths. These myths can be replaced by genuine consideration about what might be improved.

What is Passivhaus?

In the UK we use the German spelling so that it is clear what we are talking about and to avoid confusion with passive solar (see Chapter 2.5) or even passive ventilation. Also, the German *haus* refers to a wider range of buildings than the translation to *house, maison* or *casa* implies. There are Passivhaus schools, offices, health centres and archives, to mention just a few examples. You can find more details elsewhere and the UK Passivhaus Trust website is a good starting point to learn more.

Passivhaus is not a trademark and the originator, Wolfgang Feist, says it is open source. However, if someone says that a building is to Passivhaus standards then they are claiming that it actually meets the clearly documented energy and comfort requirements, whether certified or not. Unlike the meaningless terms 'eco-home' and 'sustainable building', this is a strong and verifiable claim under Trading Standards. The 2015 Passivhaus Trust document, *Claiming the Passivhaus Standard*[3] clarifies this. (This document has been translated into Spanish and adapted for North America.)

A number of buildings have claimed to be designed and built to Passivhaus principles, but this is misleading. Usually, these buildings only copy *some* of the components such as triple glazing or heat recovery ventilation but do not adhere to the actual principles, which run much deeper. If a building claims to follow Passivhaus principles and then does not, false claims are being made because the first principle of Passivhaus is that it does what it says on the tin; there is no significant performance gap.

If you want to really understand Passivhaus then it is worth visiting some buildings; kick the tyres and talk to the occupants. Some homeowners open their houses for International Passivhaus Open Days.

Like all buildings, some look great, some don't, and beauty is always in the eye of the beholder. But why do some architects say they don't like the look of Passivhaus buildings? Asking what a Passivhaus looks like is a bit like asking what a bird looks like. Even in Britain, a small brown wren does not look anything like a red kite or golden eagle and yet all are instantly recognisable as birds. This is because birds evolved within constraints, very different to those for fish or land-based mammals. There are an infinite number of ways of designing a genuinely low-energy building that is cool in summer and warm in winter but, as with flying birds, there are a number of clear constraints imposed by the laws of nature. These constraints include the building form and the size and orientation of windows and glazed doors.

A building with lots of glass that faces the setting sun will almost certainly overheat. This is due to the low afternoon sun, which cannot be shaded by simple overhangs. Such a building will fail to meet the Passivhaus standard unless it is fitted with expensive external blinds, which then block out the view. Designers who want to make buildings that feel as good as they look embrace these constraints. Those focused on external appearance see constraints as limiting free expression.

ABOVE Shepherds Barn, County Durham, by Mark Siddall LEAP 2020, an EnerPHit certified retrofit.

Passivhaus as a design constraint

The designers Charles and Ray Eames[4] saw constraints as a valuable part of the creative process that should not be confused with compromise. Charles Eames said:

Here is one of the few effective keys to the Design problem: the ability of the Designer to recognise as many of the constraints as possible; his willingness and enthusiasm for working within these constraints. Constraints of price, of size, of strength, of balance, of surface, of time, and so forth. Each problem has its own peculiar list.

Passivhaus introduces a number of clear constraints. Many of the things we love about traditional buildings were the result of constraints. When glass was handmade, panes were small. The limited availability of suitable local materials led to roofs clad in slate, clay tiles or shingles. If you were very poor then you used thatch. These constraints are mostly redundant and yet we still copy the look with fake glazing bars, artificial slate, asphalt shingles and - the most expensive - plastic thatch.

BELOW Old Holloway designed by Juraj Mikurčík using the EcoCocon prefabricated straw panel system.

OPPOSITE Ostro Passivhaus, a timber-frame self-build by Mhairi Grant of Paper Igloo, 2017.

Passivhaus myths and scepticism

Just the mention of Passivhaus seems to annoy some people. This was more of an issue when we started working on Passivhaus projects around 2008. At that time, there was little first-hand experience of Passivhaus building in the UK, but there were plenty of myths and anecdotes going round. Everyone seemed to know that you couldn't open the windows, they were cold in winter, very expensive and only really suitable for the German climate, not our milder UK weather. With false concerns circulating, many Passivhaus advocates adopted a defensive position, not daring to air any reservations or issues lest they get added to the list of problems with Passivhaus. That was 2008, and we have designed, built and visited a lot of Passivhaus buildings since then.

If buildings are to improve, we need to use an iterative approach of: design, construct, monitor and refine. One of the advantages of an international community of building professionals all working on the same problem is that we have huge potential to develop and improve. However, when people trot out the old myths it is tempting to break the monotony by sharing some actual problems.

In the UK homes and schools tend to be smaller than the equivalents in Germany and Austria. Even with a simple shape, smaller buildings have more heat loss area per unit of floor area. All else being equal, our smaller buildings require thicker insulation. The claim that Passivhaus favours bigger buildings was actually true and is true of any standard based on energy use per unit floor area rather than, say, a percentage reduction in heating energy. This was a problem and it meant that we struggled in the UK to reach the Passivhaus standard for small social rent buildings.

When we raised this problem with the PHI around 2013, they proposed a relaxation of targets for smaller buildings. However, evidence showed that these smaller homes actually used little or no heating but then tended to get too warm in summer. The answer was simple physics. While the heat loss area per square metre of floor is more for smaller buildings, so are the free internal heat gains due to occupants, appliances, cooking *etc*. The total internal heat gains in a 100m² home are generally similar to those in a 150m² home but concentrated in a smaller area. The PHPP energy model assumed a low and fixed heating gain per square metre of floor area but that distorted the reality. The two effects very neatly cancel out. Having realised this, it then took a conference paper and an article in a magazine to get changes made, but the calculations were modified to account for this effect without the need for a special dispensation for smaller buildings. (For the geeks, internal gains for dwellings changed from a fixed 2.1W/m² to 2.1+50/TFA to give a better approximation of gains as dwellings change in floor area.)

Other refinements included changes to how hot water losses are accounted for and more rigorous checking of summer overheating, again in response to local experience.

The point we want to make is that such refinement is far easier if the target is relatively fixed rather than moved further away every few years. The target remains fixed, so how we design and build can be refined to meet that target.

So, what could still be better?

Our experience has led us and most of our peers to see the Passivhaus standard as the only game in town for most new-build projects. It is possible that an even better alternative could be developed, but we don't have time to reinvent the wheel in this climate crisis where most of our buildings fail to perform as designed.

While Passivhaus is still the best methodology we know, there are changes we would like to see. Although Passivhaus is actually much simpler and more coherent than most other standards, it is still far too complex for most to understand. The standard and the associated energy model spreadsheet (PHPP) allows a lot of flexibility as to how the targets can be met. We can, for example, offset poor form with additional insulation or excess south glazing with external blinds. All this can be modelled but if we are aiming to build or renovate simple, affordable buildings at a scale that is meaningful in a climate crisis then we need simpler but no less robust models and approaches.

Interestingly, Passivhaus seems to appeal to two opposite approaches to low-energy design. On one hand, the tools appeal to minimalists with small budgets wanting to work out what really matters most. On the other hand, these tools also appeal to maximalists who can't resist the challenge of calculating all the thermal bridges and designing complex systems to manage the building as a power station.

This polarisation was well illustrated by two Passivhaus buildings featured on consecutive episodes of the UK version of the series *Grand Designs*. The first building was Juraj Mikurčik's 100m² three-bedroom dwelling, built for about £150k, and the second was a nearly 800m² concrete and steel build aiming for 'Passivhaus Premium', the PHI's take on zero carbon. That Passivhaus can embrace such a range of building types is impressive, but we need a much simpler approach if we are to make low-carbon buildings mainstream.

While Wolfgang Feist himself has described Passivhaus as 'open source' in Janet Cotterell and Adam Dadeby's *The Passivhaus Handbook*,[5] the reality is some way from that. We can build a Passivhaus and even claim that it is one without any input or accreditation from the PHI or even from local partners such as the Passivhaus Trust in the UK. That would not be possible for BREEAM or LEED, for instance, which are proprietary standards. However, it is difficult for a government or region to adopt Passivhaus without either fully collaborating with the Institute or going it alone and calling it something else. The main issue with the latter option is what gets dropped as yet another standard is added to the list?

But it's not just about energy!

Thankfully, this is not a phrase we hear so often these days – especially with the rise in current energy prices. However, Passivhaus does focus on operational energy use. A key principle is that this must not be at the cost of comfort and health, so a lot of the climate-specific requirements are about avoiding discomfort, mould and poor air quality. You can't achieve energy targets by reducing temperatures or ventilation rates, for example.

But the Passivhaus standard has no requirement to avoid materials that emit volatile

Ty Casim

organic compounds or to meet lower than regulatory levels of radon. Similarly, there is no requirement to avoid petrochemical-based insulation or to meet upfront carbon targets. But then Passivhaus has nothing to say about fire or accessibility – why should it?

Conversely there is nothing in the Passivhaus standard that stops us meeting or exceeding most other sensible environmental or safety standards. Chapters 2.4 and 2.5 demonstrated how efficient forms and optimised glazing reduce upfront carbon as well as cost. It seems strange to think that anyone who cares enough to adopt the Passivhaus standard for their buildings would not also care about upfront carbon and other environmental issues.

That said, if the building is made of steel or concrete, then the same Passivhaus performance and quality standards apply. Meanwhile, we see examples of poorly performing buildings with high energy use, justified by a life-cycle calculation that somehow credits the solid timber construction with being carbon negative. Similar arguments were applied to buildings heated by burning wood based on the now widely discredited idea that burning biomass is zero carbon.

Conflicts between Passivhaus and environmental targets

Occasionally, a conflict between other environmental targets and Passivhaus do occur. The example that comes to mind is a requirement or credit in some standards for ventilation intake and exhaust terminals to be separated by a certain distance. It seems like a self-evidently good idea to keep the air intake for our building as far from the waste air exhaust as possible. For a hospital or chip shop this is probably good advice, but for homes it is not such a good idea. Monitoring shows that there is very little recirculation of the exhaust to the inlet, even when the terminals are almost next to each other. Keeping them close minimises the length of cold duct that needs to be insulated and saves valuable space in small homes that would be taken up by additional bulky insulated duct runs. This has led to the common use of the neat combined inlet and exhaust terminal shown here. In attempting to avoid a non-problem some designers have placed the intake and exhaust on different faces of the building. As well as the longer ducts, this can lead to imbalanced flows in windy weather, which reduces efficiency and can even impact on comfort.

Another possible conflict is the requirement for higher ventilation rates for improved health. Post-occupancy monitoring of homes and schools, in particular, shows that Passivhaus buildings have consistently good year-round air quality, far better than in so-called 'naturally ventilated buildings'. However, some standards argue for higher than normal ventilation rates on the basis that you can't have too much of a good thing. In summer this may be true, but in warm buildings during very cold weather, excessive ventilation – whether from mechanical ventilation, open windows or just draughts – can result in very low relative humidity. This can create health issues, from dry eyes to increased susceptibility to airborne viral infections.

As with many of the Passivhaus requirements, it is primarily a consideration of health and comfort that drives the recommended ventilation rates, rather than energy saving.

OPPOSITE Using reclaimed Welsh slate and lime render, the north elevation of this certified Passivhaus manages to look traditional without pretending to be something it is not. Architect Adrian Cook.

ABOVE This seemingly innocuous ventilation terminal avoids cross contamination of inlet and exhaust whilst keeping ducts short and neat. See text for more details.

Passivhaus Plus and Premium

In response to the hype about net-zero buildings, the PHI developed the Passivhaus Plus and Premium categories, which include on-site generation, *i.e.* PV on the roof.

In developing Plus and Premium, the PHI took a sophisticated look at what net-zero operational energy might mean when connected to a future fully renewable electricity grid and tried to address some of our criticisms. The concept of primary energy renewable (PER) was introduced because primary energy (the amount of fossil energy needed to provide heat or electricity) or carbon is meaningless if we imagine a future without fossil fuel.

As we know, the biggest challenge with renewables is matching supply to demand. For example, solar PV is a good match for cooling but terrible for heating. Grid-scale batteries or pumped hydro might allow us to even out demand over a day, but the only practical way to provide heating during long windless spells in winter is by using synthetic gas made when supply is greater than demand and stored in salt caverns, for example. The gas is then used to generate electricity to feed into the grid. This is expensive and needs more renewable energy to make up for losses. The upshot of this is that the Passivhaus Energy model uses a different PER value for space heating than hot water or lighting, for example. In a renewable future, all energy is zero carbon and inexhaustible, but it is far from free. In the medium term, the weighting of PER according to end use also reflects the fact that we will be using fossil fuels to match demand when renewables cannot. PER as a concept attempts to deal with the paradox; if the energy source is zero carbon, how, or even why, do we set efficiency targets?

Other net zero targets simply look at total annual onsite generation and subtract

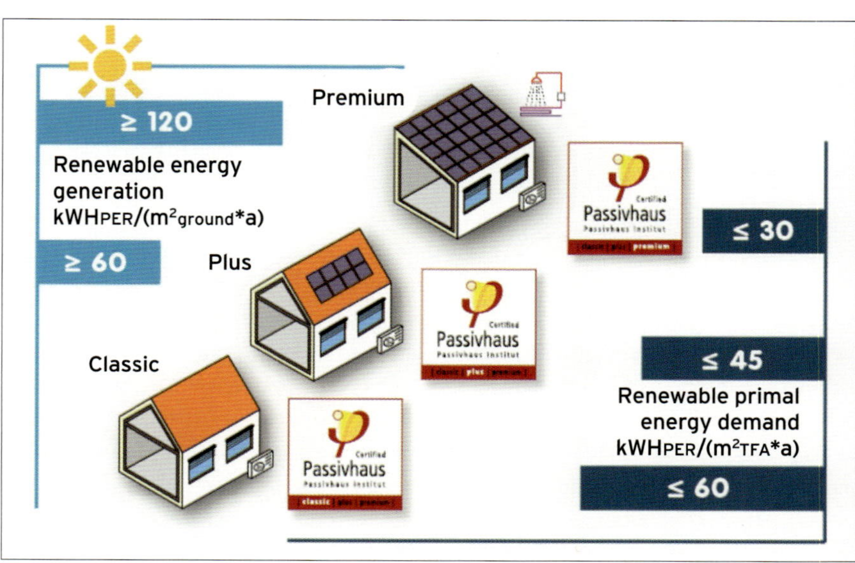

RIGHT Passivhaus Plus and Premium categories were developed as an answer to the idea of Net Zero buildings.

annual demand, ignoring the huge mismatch in timing. PV is cheap whereas inter-seasonal storage is expensive and difficult to roll out at scale – and that is why driving down the peak power and energy needed to heat our buildings is really important.

One of the other issues with net-zero building concepts is that the only realistic onsite generation option is PV. To get to conventional net zero, it is easier if the roof is relatively large compared with the floor area. Net zero rewards poor form factor with all the cost, upfront carbon and space implications.

In an attempt to at least avoid rewarding sprawling low-rise buildings, Plus and Premium instead reward the proportion of available footprint devoted to PV, allowing even a tower block to theoretically achieve the standard, even though the on-site generation will only provide a small proportion of demand.

What about retrofit?

For new build, we are confident that most building types should be built to the Passivhaus standard or equivalent. We know it works and we now have examples of Passivhaus buildings constructed for no extra cost. If we all do it, we will see costs come down and quality will improve for everybody. We need to stop making buildings that will immediately need retrofitting once legislation that is currently held back by vested interests finally catches up.

Retrofit is an altogether more difficult problem. Unsurprisingly, Passivhaus has a solution: the EnerPHit Standard. You can find details of the standard and numerous case studies in Marion Baeli's book.[5] The logic is that if windows need replacing, for instance, then this provides the perfect opportunity to upgrade to triple glazing, and the extra cost is quite small. If we miss that opportunity and install less-efficient products, we will be locked into lower performance for another 30 years or more. Similarly, if we need to renew the render on our walls, it is a great opportunity to add some insulation at the same time, as

BELOW The Entopia Building is a former telephone exchange that has been renovated to the Passivhaus EnerPHit standard but also meets BREEAM Outstanding and Well Gold standards. Design by Architype and BDP.

the additional cost of the insulation is relatively small and the benefits in terms of comfort and eliminating mould are high. Once we have decided to insulate, the extra cost of adding a little more is quite small and far less than the unlikely event of adding more to a wall that has already been insulated. And so it goes on – if the roof needs redoing why not get the roof or ceiling insulation to join up with the wall insulation?

Any suboptimal refurbishment means we are locked into a long period of ongoing high-energy use. As with new builds, using major renovations as an opportunity to future-proof the building makes economic sense – or rather it makes less sense *not* to do it. Thus the accepted wisdom in the Passivhaus and low-energy building world has been that we should only do deep retrofit to the Passivhaus standard (EnerPHit) or equivalent. Anything less locks us into higher lifetime emissions.

The problem is that we need to decarbonise now. Relying on rapid deep retrofit is hard to imagine given the cost and complexity, not to mention the sheer hassle of major building work on occupied buildings. An additional consideration is that all works emit CO_2 so deep retrofit comes with an increase in upfront carbon.

Heat pumps

The easiest solution, particularly as the grid decarbonises, is to roll out air source heat pumps. Here the conventional wisdom was that unless we greatly reduce the heating demand, the heat pumps will not work and will have to use very expensive direct electric heat to boost output in cold weather. In turn, this could mean that the grid will fail to meet the increased peak electricity demand in winter.

There are plenty of anecdotes of heat pumps being installed leading to high electricity bills and poor performance. However, a growing number of practitioners are questioning

LEFT This low budget self-build renovation in Cheshire achieves the full Passivhaus standard. The owner trained as a Passivhaus consultant and managed the build with architectural design by Gil Shalom of GSD Architecture.

ABOVE Glyn Hudson of Open Energy Monitor achieved excellent comfort and low bills for his parent's 1980s bungalow by optimising the radiator sizing and control but without fabric upgrades.

the conventional wisdom and managing to achieve surprisingly good results in houses that have had minimal efficiency measures undertaken.

In a recent guidance note for the UK Passivhaus Trust, Alan Clarke and Kate de Selincourt[7] looked at how optimised heat pump installations with basic efficiency improvements can provide improved comfort and reduced carbon emissions in existing homes without difficult-to-implement wall insulation. They found that if radiators are well sized and the heating is run continuously in winter, the heat pump can run at a low temperature, improving the COP enough to compensate for the additional heat input. By running the heating continuously with only a small temperature reduction at night, the comfort is improved and risk of mould is reduced.

However, it is not a simple case of just bolting on a heat pump. Any draught proofing must be complemented by effective and continuous mechanical ventilation and users must embrace the counterintuitive approach of continuous heating. Those on a pay-as-you-go meter and with limited means will be tempted to heat intermittently and will not enjoy the comfort and health benefit of steady heat input.

Why retrofit?

When considering retrofit – whether on a national scale or for a single building – it is useful to be clear why we are doing it, as the appropriate solution will depend on what we are trying to achieve.

If *decarbonisation* is the main driver then fit a heat pump now. The heat pump may cost more to run than a gas boiler but the carbon savings will be significant, as Richard Erskine[8] asserted in his 2022 blog.

If *comfort and health* is the primary driver, then we need to tackle ventilation and cold surfaces such as windows and walls. This should save energy but that is not guaranteed as we may choose to heat the building more once that becomes affordable – a version of the rebound effect. This is one of the arguments for the need for deep retrofit if we are to deliver actual energy savings.

If *cost* is our primary driver, then we will probably choose to sacrifice health and comfort. If we can't afford heating, we are very unlikely to be able to afford to carry out a deep retrofit. Such measures need a big picture view and social funding. The first low-energy lightbulbs cost more than ten times what they cost now, yet still provided a good return on investment – better than savings accounts at the time. They take seconds to fit and yet uptake was limited without subsidies or giveaways.

If you want to save money and decarbonise but don't have capital to spend then downsizing or increasing building occupancy is something to consider. If you are a couple of empty nesters rattling around in a large family house, this could be a great option, but it is obviously not viable if you are at maximum occupancy in cheap rented accommodation.

Is Passivhaus always the solution?

Although originally developed as a solution for housing, Passivhaus has been applied to most building types you might think of. Perhaps surprisingly, the same energy targets applied to schools and offices are also applied to single family homes. It is only when we look at supermarkets, swimming pools or hospitals, for example, that serious reappraisal of the targets and approach needs to be made.

In Chapter 1.2 we looked at the arguably niche world of museum and archive buildings. There are certified Passivhaus archives, but we suggested that this may not always be the best way to go as human comfort and ventilation are not considerations in most archive stores.

So while Passivhaus is not perfect, it is undeniable that the vast majority of Passivhaus buildings perform as designed in terms of energy use and comfort. If a Passivhaus building fails to perform we can find out why and fix it - it is not a mystery or something to blame on the occupants. As Wolfgang Feist observed all those years ago, buildings must obey the laws of physics.

BELOW St Sidwell's Point leisure centre, Exeter, 2022. Lead architect Space and Place with Passivhaus building envelope design by Gale and Snowden architects.

Buildings must breathe

In most climates, a good house is draught free but properly ventilated. This is good for comfort, good for saving money, good for reducing CO_2 emissions and good for the durability of the building. So why is airtightness such an emotive subject, evoking statements such as 'I couldn't live in a hermetically sealed box'? This phrase shows a surprisingly widespread and persistent confusion between airtightness and ventilation. A leaky building does not guarantee good ventilation. When it is cold and windy, we get uncomfortable draughts.

The usual response is to try and stop these by closing any vents and perhaps sealing up draughty windows for winter. If we manage to stop the draughts then when the wind drops, there will be insufficient ventilation to remove excess moisture and odours. The humidity will rise, particularly in cooler rooms such as bedrooms, and mould and dust mites will thrive.

However, if we crack a window open in the bedrooms there is no guarantee that this will help. In fact, this can make things worse. Typically, bedrooms are upstairs so they can become the route for stale moist air leaving the building by the stack effect. This is when the warmer – and so less dense – indoor air tries to take off like a hot air balloon but leaving the heavy, leaky house behind. Our open window becomes an exhaust rather than an inlet for fresh air, funnelling the air moistened and polluted by cooking and bathing through our bedrooms, raising the relative humidity and increasing the risk of moulds.

A tight building is a durable building

In winter, warm room air with a high moisture content leaks into the insulated wall and roof construction of a leaky building. As this air passes through the insulation to the outside, the temperature drops and condensation forms, potentially leading to mould and rot. Vapour barriers are specified in an attempt to prevent this *interstitial condensation* and calculations of vapour movement are performed for building warranties. But it is bulk air leakage rather than water vapour diffusing slowly through solid surfaces that causes most of this condensation. When the wind changes direction, air leaks back into the building, potentially carrying mould spores and insulation fibres – natural but unhealthy.

Breathability is a misleading term as it implies air movement – like when we breathe. To describe a building element such as a wall or roof as breathable means that water vapour can escape rather than building up and potentially causing rot. Think breathable waterproof fabric that protects against wind and rain but also reduces the build-up of perspiration. A woollen jumper is also breathable but offers no protection from wind and rain. In a heating climate, the rule of thumb for an element such as a wall or roof to be deemed to be breathable is that the cold side should be at least five times as open to water vapour as the warm side. One way to ensure this is to use a completely vapour-tight layer, such as a polythene or aluminium foil barrier on the warm side. One risk with this approach is that in hot weather, vapour could be driven in towards the cooler internal face where it would be trapped and condensation could occur. This is more of a concern with retrofitted internal wall insulation where any moisture must dry into the building. Another concern is that rain ingress during construction can be trapped within the building envelope and will be unable to dry out, causing rot or rust.

Measuring airtightness

Building envelope air leakage can be measured quite easily. It is done with a purpose-made fan that fits in a door or window, allowing the building to be pressurised or depressurised. The airflow can be measured and the leakage is calculated at a standard pressure of 50Pa. The average leakage over the heating season is about 5 to 10% of this figure, depending on exposure, and this allows the energy loss due to infiltration to be calculated.

Because airtightness can be measured, there is no hiding of poorly planned details or workmanship behind plasterboard or trim. This has led to improvements in quality beyond the lack of draughts and energy saving with reduced defects. For the first time, we can put an independently verifiable number on an important aspect of build quality.

Design and build tight

The Passivhaus standard raised the bar by requiring a Passivhaus airtightness (n50) of less than 0.6 air change rate (ACH), which is about 10 times higher than standard UK buildings. At first this seemed to be a huge challenge but now we regularly see builders achieving less than 0.2 and even some first-time builders achieving under 0.1. The sense of pride is heartening as teams strive for better and better scores.

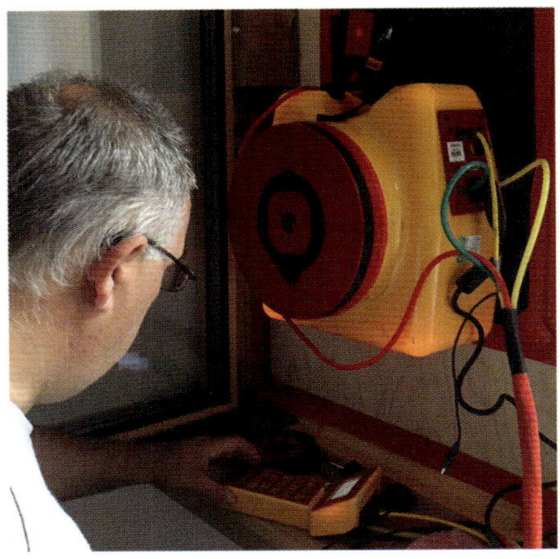

TOP Exterior of Tanzanian eco-village building.

ABOVE Paul Jennings has been testing buildings for airtightness for over 30 years. This small fan with a tiny orifice plate is all that is required to pressurise a Passivhaus home to the 50 Pa test pressure.

If we start with a poor airtightness target and apply incremental improvements then we find that the effort required increases as we aim for tighter and tighter targets. We see this when improving existing buildings. There will be some easy gains – big holes that can be quickly filled, but as the building gets tighter we spend more and more effort for smaller and smaller gains. The law of diminishing returns is what we expect and some would argue that we should aim for a middle value rather than striving for perfection.

Paradoxically, if we jump straight to a tenfold improvement in airtightness (or anything else) we have to take a radically different approach. We have to design out the leaks rather than spending lots of time and materials trying to seal them. Taping around rafter ends or floor joists is mind-numbing, time consuming and expensive, so don't do it!

Good airtightness saves energy and reduces heating costs but turns out to be particularly important for reducing peak heating load in cold and windy weather. This is something Jon experienced first-hand when his boiler failed in January. Designed and built before we fully understood the importance of airtightness and thermal bypass, his house is well insulated with cross-battened stud walls to reduce thermal bridging. In cold still weather, it is much easier to heat than when the wind gets up. Many experienced practitioners designing to the Passivhaus standard are choosing to design to the 10W/m² peak heating load target rather than the more usual 15kWh/(m²a) heating energy target. Average heat loss due to infiltration depends on average air leakage, but for peak load calculations we assume windier conditions. The PHPP peak load calculation multiplies infiltration by a factor of 2.5, so we find that improved airtightness is rightly rewarded. Once we have designed airtightness in, it is essentially achieved for free.

ABOVE Achieving economical airtight construction is 90% design. Here it was 90% effort on site.

LEFT Airtight OSB with no structural penetrations gave builder Dai Rees an n50 of 0.07ac/h at the preliminary test, almost 10 times tighter than Passivhaus requires and around 100 times better than current building regulations.

HOW AIRTIGHTNESS IS MEASURED
--

The units used in the UK are: cubic metres of air per hour per square metre of building external area (walls, floor, roof) at pressure difference between inside and outside of 50 Pascals.

An alternative metric currently used for Passivhaus is: air changes per hour at 50Pa or n50. For info: n50 if air changes per hour at 50 Pascals.

For a typical house, the numbers are almost interchangeable for larger or smaller buildings – the two metrics diverge considerably.

Ventilate right

Natural ventilation conjures up lovely images of a summer breeze wafting through meadows and rustling leaves, whereas *mechanical* ventilation conjures up images of noisy fans and high energy consumption. For most climates, including the UK, we advocate the latter. The journalist Kate de Selincourt has said that relying on so-called natural ventilation to provide good indoor air quality is like throwing food at your kids and expecting it to provide them with a balanced diet.

As with airtightness, this is another surprisingly emotive subject. All proper controlled ventilation benefits from excellent building airtightness. If the building is full of holes then it will be impossible to regulate the ventilation, which will be largely driven by wind.

The authors would favour a simple passive approach to ventilation but we have not managed to make this work. We are not saying it is impossible but it seems to be very difficult to achieve controlled ventilation by passive means under the full range of weather and internal conditions experienced in most buildings. Most of the problems can be solved by using a small fan to provide controlled ventilation whatever the weather. With good design and efficient fans, the power required is around 5W per person or less. So, we have embraced mechanical ventilation in various forms.

People say that you can't open the windows in a mechanically ventilated building. The truth is you can open the windows but, more importantly, you can also close the windows without air quality suffering. Like Einstein, we suggest that things should be as simple as possible, but no simpler.

RIGHT Alan Clark commissioning the ventilation in a small Passivhaus using a flow finder hood.

LEFT Cartoon by architect Louis Hellman.

Mechanical ventilation

Everyone is familiar with the noisy fans in bathrooms that come on with the light switch and run on when the light is off. In UK building regulations, this is called a Type 1 system and we do not recommend it. It is much better to have a continually running but very quiet fan extracting from toilets, kitchen and bathrooms. A boost switch is useful for cooking and times of higher occupancy but turns out to be unnecessary for bathrooms. Each of these rooms can have a separate fan, but a central fan with ducts is quieter and more efficient. Fresh air is admitted to bedrooms and living spaces via vents that usually include passive control that opens and closes a flap in response to the relative humidity in the room.

The next level of sophistication is to balance supply and extract with fresh air blown reliably into living rooms and bedrooms. This allows fresh air to be filtered, removing pollen and particulates, and makes heat recovery possible but also moisture recovery in dry or humid climates. In the UK, we usually call this MVHR, but comfort ventilation sounds more appealing.

The downside is an extra fan for supplying fresh air, two filters to keep the heat exchanger clean and extra ducting to get the fresh air to living and sleeping rooms. As we can recover more than 90% of the heat, we can ventilate in cold weather without discomfort or high heating consumption.

EcoCocon's new straw panel factory in Voderady, Slovak Republic, by Createrra just before the robotic assembly lines were installed.

3.0 Low impact construction

3.1 Heavyweight or lightweight? Advantages and disadvantages of both approaches and some myths explored

3.2 Upfront carbon emissions How to reduce emissions arising from the act of construction, alongside other environmental impacts

3.3 The more details, the more devils Principles for detailing to create robust buildings that achieve design, performance and sustainability aims

3.4 Elements of good building Approaches to the design of foundations, raised ground-floor construction, non-loadbearing walls, flat roof construction and windows

3.5 Services Efficient building services that can be integrated into the design from an early stage

Heavyweight or lightweight?

The question of whether buildings should be heavyweight or lightweight was one of the many topics that split opinion when green builders and designers got together. Some practitioners were building earth sheltered concrete bunkers while others favoured timber construction on posts with lightweight cladding. Thermal mass is a well-understood physical property of materials, so why the huge range of opinion and uncertainty? Why did the discussions seem more like religious debate with little chance of resolution rather than scientific exploration of a problem with a solution?

It is not that calculations and modelling were not done; both schools of thought would reference dynamic models. In hindsight, the fact that the proposed solutions covered such an extreme range is an indication that perhaps thermal mass is actually not such a big issue. This discussion also overlaps with the debate about passive solar design. Mass and glass go together like getting drunk and hangover cures.

Advocates of heavyweight buildings point to mass as a thermal battery soaking up daytime gains to keep the building warm overnight. This is one of the reasons why our early ancestors occupied caves, which average out the annual temperature cycle. Meanwhile, enthusiasts for lightweight construction point to the fast temperature response with quick heat up from cold but also faster cool down when windows are opened at night. Examples of this approach have existed since pre-history. They also point out that lightweight buildings tend to use renewable materials such as timber rather than high environmental impact concrete. The heavyweights will point to the low upfront energy of Earthships made from earth rammed in old tyres.[1] We have been around the debate for so long we could happily make a convincing case for either camp.

Mud
Kitchen

Thermal mass and insulation cannot be seen in isolation. The combination of how much heat can be stored for a given temperature rise (thermal mass) and the rate of heat loss for a given temperature difference between inside and out (heat loss coefficient) sets what is called the time constant of the building. This is a measure of how fast the building will heat up with net gains or cool down due to net losses.

Up to a point, you could get the same time constant with a lot of mass and less insulation or a lot of insulation and less mass. We might experience an uninsulated massive building retaining heat and staying warm as the outside temperature drops at night. But only insulation stops actual net heat flow from the warmer side to the cooler. In a few rare climates with high daytime temperatures and cold nights we can achieve reasonable comfort with mass alone by simply averaging the night and day temperature plus some solar and internal gains. However, for most of the planet we need to keep heat in or out to maintain comfort.

Heavy insulation

A debate that also goes on relates to the heat capacity of the actual insulation. If you do an internet search for 'decrement delay' you will find very convincing arguments for why heavy insulation saves more energy and reduces overheating much better than lighter insulation of the same U-value. This may feel right, but the science does not support it (see the Passipedia article, 'Insulation vs. Thermal Mass'[2]).

We think this apparently unresolved debate was due to the very anecdotal nature of much building performance data with lots of unknowns including significant gains from inefficient lights and appliances, air leakage, thermal bridges, thermal bypass, random uncontrolled ventilation and unmetered sources of heat, such as from woodstoves for top-up heating. Over the course of our working lives we have seen a lot of these uncertainties addressed as practitioners have tried to close the performance gap between design and reality. In the UK we were somewhat behind the curve.

Heavyweight and lightweight: neither good nor bad

What we now know - and others have known all along - is that while thermal mass will influence how a building works, it is neither good nor bad. We have good examples of heavy and lightweight buildings all meeting the same Passivhaus targets for energy use and summer comfort. The heavyweight buildings can in theory tolerate slightly higher daytime gains in summer from occupants and sun, but unless that stored heat is purged overnight the building will slowly heat up. Having overheated, it might take days to cool the building down again.

In the UK we famously have weather rather than climate and we will generally welcome the onset of a heatwave and the higher building temperature, knowing it will probably only last three days. But by the time we realise we are in for a headline-making scorcher it is too late to dump the heat and night-time temperatures may well have risen so that night cooling is not an option.

Another issue is that an idealised dynamic model of a heavyweight building with

OPPOSITE Post-Occupancy Evaluation of Ysgol Trimsaran shows excellent summer comfort despite the lightweight timber construction.

BELOW Scatter plot of indoor temperature (half-hourly mean) against outdoor temperature (half-hourly mean) in the non-heating seasons in Ysgol Trimsaran, from monitoring in the school year 2018-19. The red lines suggest the temperature range for human comfort which also depends on historic outside temperature.

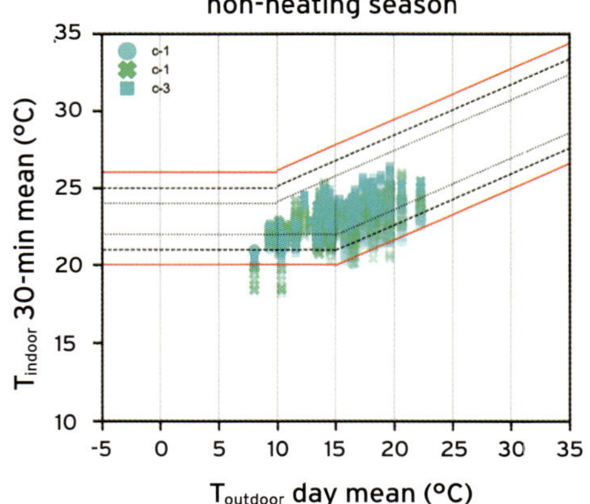

perfectly controlled night ventilation is likely to credit the benefit of thermal mass hiding any problems due to excess glazing. Indeed, any modeller worth their salt tasked with proving compliance for building regulations will know which knobs to tweak to achieve a pass without the awkward conversation of suggesting any changes to the fenestration.

Conversely, without the assumed safety net of additional thermal mass, the lightweight building will struggle to meet theoretical compliance, leading to design changes to glazing area and shading. As the thermal mass benefit will only work if the building is operated correctly, the heavyweight design is likely to be less robust in reality.

Schools

Schools are one of the building types that might benefit most from higher thermal mass to soak up the gains from kids and sun in the day. In practice, lightweight timber schools work fine with just the plasterboard as added thermal mass. In our experience, summer discomfort tends to correlate with excessive glazing rather than lack of thermal mass.

Sharon Owen,[3] headteacher at Ysgol Trimsaran commented, 'It is spacious, light and airy and has improved pupil's pride and self-esteem. The temperature is constant thus improving concentration levels.'

The other argument in favour of thermal mass is that it stores free solar heat. We covered the high cost of additional glazing as a heat source in Chapter 2.5, but even normal levels of glazing will introduce gains that will be useful at night. The PHPP lets us model this and we find that the benefits due to increased mass are very small. This is borne out in practice with no evidence that more massive Passivhaus buildings use significantly less energy for heating than lightweight ones. If there is an effect then other variables mask it.

Reasons for heavy construction

In practice, we find the choice of heavyweight or lightweight construction are driven by other factors. Hope View House , which we described in the passive solar chapter, is dug into a slope for planning and architectural reasons. A timber construction would have been possible but here masonry was the simplest and most robust solution.

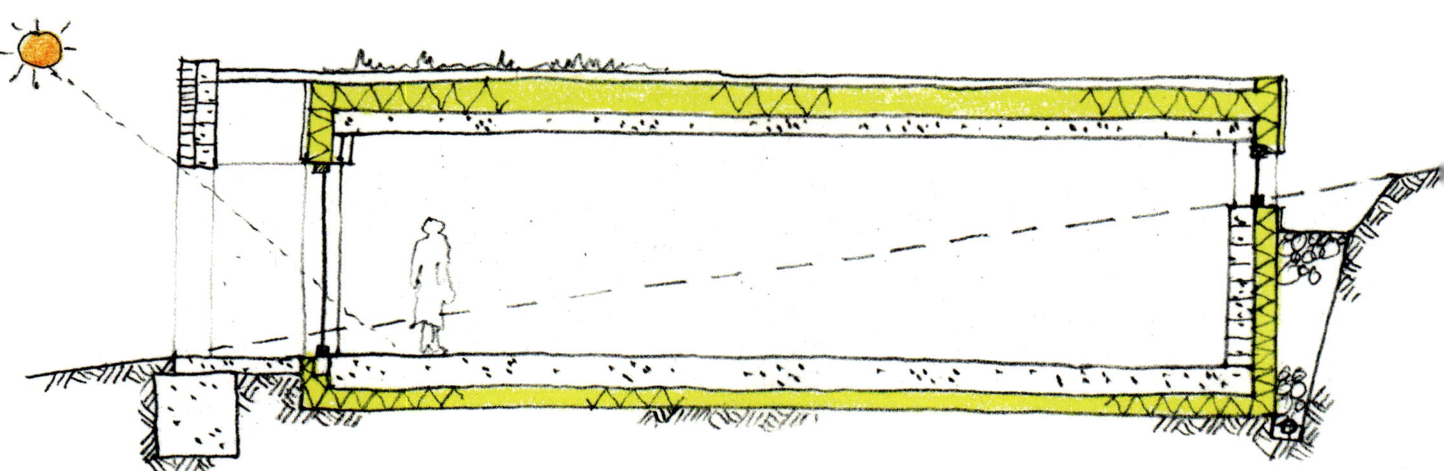

BELOW Section: Hope View House is built into the hillside to the north and has a deep overhang to glazing on the south. Masonry construction was chosen for structural reasons. Warren Benbow Architects 2017.

Heavyweight	Lightweight
Can deliver excellent comfort and low energy use.	Can deliver excellent comfort and low energy use.
Can soak up gains in the day, keeping temperatures down, as long as that heat is purged overnight.	Will heat up and cool down more quickly. Ideal if the building has been unoccupied for a few days.
Tends to mean concrete for mainstream buildings.	Timber is favoured but light steel frame and other materials are possible.
Typically used for very secure buildings such as archives.	Secure, fire-resistant, lightweight options are available.
Beware misleading claims about mass acting like insulation to reduce heat loss.	Ground-bearing floor slabs provide some benefit in summer, even if insulated. Unlike massive above-ground fabric, the ground will not get too warm.
Designs reliant on mass and glass are very sensitive to failure due to heatwaves and user understanding.	Some thermal mass is useful to reduce diurnal swings, typically the plasterboard needed for fire and acoustic performance.
Thermal mass may become less available if floor coverings or acoustic ceilings are added.	
Favoured by concrete industry and heavyweight natural material enthusiasts (rammed earth, cob, hempcrete).	Favoured by steel and timber industry and natural materials enthusiasts working with wood, fibre insulation and straw bales.
Buildings in the ground or against retaining walls for structure and moisture resilience will usually use massive materials such as masonry and concrete.	Timber basements are possible but not common outside of North America.

ABOVE Lightweight construction in the tropics where raising the building creates shade in the undercroft and breeze upstairs to moderate the heat and humidity. These self-build houses are in Guyana where the national housing policy is providing grant aid for people building their own houses.

Upfront carbon emissions

We feel confident writing about reducing operational energy in buildings because we now have a pretty good idea of how to do that and first-hand evidence of the results. We are less confident writing about the even more important but slippery topic of upfront emissions - those due to manufacture and construction. These are often called embodied energy or carbon, but that is misleading, especially when discussing bio-materials such as timber.

Embodied carbon could mean carbon taken out of the atmosphere and locked up in a building for many years but it also refers to all the emissions that end up in the atmosphere due to construction, plus we can't avoid considering end-of-life emissions. Not only is it complicated, there is a lot of what is currently taken as accepted wisdom that we are confident is demonstrably not true. Note: as we are not quoting numbers in this chapter, we will use the words carbon, emissions and carbon dioxide loosely to refer to any emissions that will increase global heating or even temporary cooling (carbon particles, say).

We are genuinely baffled by the various metrics for upfront carbon targets that we come across in our work but we do want to expose some naked emperors that continue to be praised for their attire. We can point to some obvious but usually ignored ways to genuinely reduce upfront impacts. A very readable primer on this topic is former architect Lloyd Alter's book, *The Story of Upfront Carbon*[1].

In Chapter 1.4, we had a sceptical look at a self-evidently ultra-low-carbon exemplar: a house made almost entirely of renewable cork. In this chapter we will call out what we see as some creative carbon accounting. The absolute carbon reduction of a design decision may be hard to quantify in terms of kilograms of carbon dioxide equivalent (CO_2e) emitted, but if we use half the amount of something we can be confident about the relative reduction. That said, we should not fall into the trap of banking credit for a reduction due to not doing something! We need to make drastic reductions on net global carbon emissions and can't credit polluting materials we claim to avoid. If we are trying to reduce our calorie intake, all that matters is what we actually eat, not how many all you can eat buffets we walk past - welcome to the world of carbon accounting. To be clear, avoiding building does avoid emissions in the same was as not eating a cheesecake does avoid calories - we just don't think you can use that credit to offset another building or another slice of cheesecake.

How to measure upfront carbon

TERMINOLOGY FOR UPFRONT EMISSIONS

--

Terms and concepts such as cradle-to-gate are giving way to cradle-to-grave and cradle-to-cradle:

- Cradle-to-*gate* only includes emissions from extraction to leaving the factory gate. It excludes transport to site, use and disposal because these variables are not usually under the control of the supplier.

- Cradle-to-*grave* includes the whole life cycle of a product, including assumptions about end-of-life disposal.

- Cradle-to-*cradle* products are based on the idea of a circular economy, whereby materials and components can be reused or recycled indefinitely.

Upfront emissions are more difficult to model than operational ones and much trickier to measure. For operational energy we have a clear boundary, the building, and one or perhaps two consumer meters recording the energy used. It is not much more difficult to widen our boundary to include where the energy comes from so that we can calculate the carbon emissions and peak load on the grid, including any match with renewable generation. The upfront emissions associated with construction are much harder to measure. For a start, where do we draw our boundary?

All of these attempts at carbon accounting still have to draw an arbitrary boundary if they are to avoid trying to calculate the almost endless carbon trail that exists for any building material, component or activity. Clearly, we should include the energy and emissions to transport a product – particularly if it is bulky, like insulation, or heavy, like stone. But what about including a share of the life cycle emissions of the truck and the driver? A recent article by Candace Pearson and Elizabeth Waters[2] claims that construction site emissions, previously thought to be relatively insignificant, could account for 'as much as 30% of upfront embodied carbon'. We have no way of easily checking such claims and that is the problem.

And how was the money earned to pay for the materials? How will the wages and profits be spent? This starts to matter more if we are paying a premium for a low-carbon product. Some make the case for bigger budgets to reduce emissions but even abstract money exchange has carbon implications. People with bigger brains are working on all this, but it feels like we have some way to go before we have the same level of certainty that we do for operational emissions.

RIGHT This house in Devon was self-built by Kevin McCabe using soil from the site, straw from the local farmer and local oak and chestnut for the floor and roof structure and for the joinery. It is a very low upfront-carbon building but suffers from shrinkage causing air leakage.

Two schools of eco-building

Among eco-building pioneers – as with heavy versus lightweight and passive solar versus superinsulation – the debate has tended to be polarised. On one side we had those prioritising the reduction in upfront energy, usually at the cost of operational energy use. On the other side we have those prioritising reducing operational energy above reducing upfront energy. At the extreme, the earth building enthusiasts are digging clay from the ground on site, perhaps mixing it with local straw and building using a vegan workforce who cycle to site. This seems to be about as sustainable as we can get and is not quite as niche as it may sound. The hardcore advocates will be skip-diving for temporarily out-of-fashion bathroom suites and making their own secondary glazed windows. We applaud these people, but our concern here is the mainstream. Unfortunately, despite some claims to the contrary, mud is not a great insulator and we will struggle to meet even building regulations levels of energy efficiency without serious upgrading of all other elements.

The problematic argument that is sometimes made is that by avoiding upfront carbon from conventional materials, we can afford to emit more energy to heat the building. Typically, such buildings were heated by burning wood, which was thought by many to be almost carbon neutral. This meant that adding wall insulation, heat recovery ventilation or triple glazing would actually seem to increase the life cycle emissions. It is now generally accepted that burning wood or pellets emits more CO_2 at the chimney than burning coal. Sure, if we grow more trees the CO_2 emitted over the full life cycle will be close to zero, but if we hadn't burnt the trees there would be a lot less CO_2 in the atmosphere.

Nick built his house based on a belief that wood was a zero-carbon fuel. He still cared about insulation because it would save the time and cost of cutting firewood. In wrestling with the paradox of why to save energy if the energy is zero carbon, in 2010 he and Alan Clarke wrote a short discussion paper, 'Biomass – A Burning Issue',[3] which caused a lot of controversy and upset at the time.

We could, of course, replace the smoky wood burner with a modern heat pump powered by an ever lower carbon grid. Now imagine earth and straw panels manufactured in a modern state-of-the-art solar-powered factory and this argument is not obviously wrong.

At the other extreme are those advocating high-performance petroleum-based insulation, perhaps bonded into factory-made panels. The argument being that over a 60-year life cycle, it is the operational rather than the embodied energy that dominates. As we move to a decarbonised grid and heat pumps, we need to dramatically reduce peak heating load if we are to meet the demand without a huge increase in costly and carbon-intensive infrastructure. Also, if we don't take a hit now on upfront emissions to reduce demand then we are locked into a future of high operational energy use – a convincing argument.

Some have simply seen this as a false dilemma. For example, the EcoCocon straw wall panel system is both thermally efficient and has low upfront emissions. It is certified as a cradle-to-cradle product and as a Passivhaus building component. But walls are only part

ABOVE The lightweight approach.

of the building. What about high-performance windows and doors, heat pumps, roof coverings, plumbing, bathroom suites and flooring … the list goes on.

The third case to consider is similar to the apparently extreme earth building example: existing buildings in need of refurbishment. The construction emissions have already happened, so we are starting from zero. What is the right balance between potentially huge carbon emissions now to retrofit buildings to a high standard versus higher emissions over the life of the building if we stop short of a deep energy-efficient retrofit? We explore this dilemma in Chapter 2.6.

The zero-carbon building myth

Faced with political pressure to commit to building new housing and a conflicting commitment to carbon reduction targets, someone came up with the fat-free-cheesecake idea of *zero-carbon buildings*. The idea that buildings could have no net impact is clearly appealing but sadly untrue.

We need to embrace net zero at a global level so that sequestration on land and in oceans is greater than global emissions from humans and the rest of nature. We need CO_2 levels to stabilise and ideally fall. However, as everything we do causes emissions, that requires sequestration.

The problem with sequestration is similar to the problem with credit. Used wisely it is a valuable tool to balance cashflow. The problems happen when we see it as income. The carbon books must be balanced globally and there is a very limited capacity for sequestration, which also has to offset natural carbon emissions that would happen without us.

But sequestration gets sold as a commodity to offset our CO_2 emissions without changing our consumption – this simply doesn't add up. We are told we can offset any emissions by planting some trees and we don't even have to plant them ourselves. Thanks to a global market with cheap land and labour and an urgent need for foreign revenue, the cost of such offsetting is remarkably small. A quick internet search suggests I can offset a tonne of carbon dioxide equivalent (CO_2e) for the price of a pint of beer. I also stumbled across travel firms offering zero-carbon ski holidays, including flights. We really want to believe this is true.

The unavoidable issue is that we can't grow enough trees; indeed, we are losing forests globally. We strongly advocate timber as a building material but are painfully aware that construction timber is in increasingly short supply.

It gets worse. The carbon offset start-ups are realising that planting and nurturing trees to maturity is a right pain! The risks from wildlife, drought, flood, gales and disease are only made worse when we plant a monocrop for short-term profit. We remember the catchy slogans, 'plant a tree in '73; plant some more in '74 [...] the hopeful still alive in '75', only to be followed by 'dead as sticks in '76'! More recently, planting hundreds of ash trees only to watch the leaves start to turn black as dieback takes hold.

It turns out that a far better business idea is to simply pay owners of pristine forest not to cut down their trees that are already standing. Before actually writing that sentence, we had to do an internet search to confirm this is a real thing used by the likes of Microsoft and Royal Dutch Shell, not something we'd misheard down the pub!

Aside from the very real issue of long-term auditing of such schemes, it is difficult to know where to start unpicking this creative approach to carbon accounting; what else can we pay people *not* to do, so that we can then do that thing ourselves?

While the nonsense of these schemes is being exposed, variations of the above are what allow companies and organisations, including local councils, to claim what should be incredulous progress towards being zero carbon in under 10 years, usually without actually doing anything more than paying for offsets and installing some optional cycle racks for the photo opportunity. Indeed, offsetting is currently so cheap that companies are offering zero-carbon electricity and gas for no extra cost.

As with any other Ponzi scheme, the sums can be made to work if we choose our boundaries and time scales carefully. Our carbon offset market is at the early rapid growth phase where supply outstrips demand. Even though actual global emissions are still increasing, a growing number of annual reports show emissions of those companies or councils rapidly heading towards zero and for little or no cost to shareholders or council tax payers. Looking ahead, this elegant free-market solution starts to quickly fall apart in too many ways for us to fully unpick in a book about building.

Capture and storage

We have to at least mention carbon capture and storage. At the time of writing this does seem to be an expensive PR stunt, but even if it can be made economical and be scaled it will most likely be used to justify the ongoing burning of fossil fuel.

At the building level we see yet more smoke and mirrors and creative accounting. A common myth is that if we use enough timber in our building this can result in a building that removes CO_2 from the atmosphere. The 2021 Serpentine Pavilion was claimed to have removed a net 31 tonnes of carbon from the atmosphere, despite containing 95m^3 of concrete and a lot of steel. This was due to the timber, cork and plywood *locking up carbon* in this temporary structure. Architect Stephen Parnell was quick to suggest that all we need to do is build more pavilions and we can sort runaway climate change![4] He was joking and I really wish I didn't have to spell that out.

RIGHT The built environment accounts for 14.4 metric gigatons of carbon dioxide equivalent (GtCO2e) of emissions around the world annually so we need to build 465 million magic negative emissions pavilions a year to offset built environment emissions.

Even buildings made from annual crops such as straw, which can be claimed to actually lock up carbon that would have returned to the atmosphere, still result in carbon emissions. Using renewable bio-materials should result in lower upfront carbon emissions than using steel, concrete and plastics, but we can't build our way out of climate change.

Looking forward to a more rigorous methodology for upfront carbon accounting that closes all the loopholes and gaming of the system, we still need to consider design for durability, adaption and reuse. While most of the life-cycle calculations assume a rather brief 60-year design life, a good number of modern buildings need to be demolished after a much shorter life. Well-detailed timber and straw buildings can last for hundreds of years but there are a worrying number of flagship low-carbon buildings that have failed catastrophically after a very short period. It would be unfair here to pick on examples. We see such failures as design problems rather than material problems and while we are happy to specify timber in buildings we are aware that under the right conditions, organic materials will rot very quickly.

Reuse and recycle

All new building is going to be damaging to some extent in strictly environmental terms and so you should only build new when all possibilities of refurbishing and reusing existing buildings have been exhausted. Prefabricated timber buildings are routinely taken down and reused in North America and elsewhere. In Britain, a number of Segal Method buildings have been taken down and the materials reused to build new buildings including an office moved from Waterloo to build a nursery in Stockwell in London, and a demonstration house built by volunteers at the Glasgow Garden Festival and re-erected in the east of Scotland to a different plan.

Designing to combine materials and components in their manufactured sizes as far as possible minimises waste, as does employing prefabrication where waste tends to be recycled in the factory rather than dumped on site. The site treatment of contaminated ground and the efficient design of foundations removes the need to take spoil to landfill, which just moves the problem at a substantial cost.

Limits on the use of reclaimed materials

There are a number of issues that act as disincentives to the growth in the use of reclaimed materials (that is, materials that can be reused with little or no processing; second-hand bricks, for example). The market for second-hand materials is underdeveloped in Britain and is restrained by the common perception that they are inferior. In practice, the reverse can be the case. Reclaimed timber is generally well-seasoned timber cut from mature trees with close, straight grain and few knots. Second-hand bricks and tiles are often handmade with a depth of character and texture that a machine-made article can never attain. Also, the market is limited by the general reliance on standards, quality assurance and guarantees, all of which are often difficult to obtain for second-hand materials.

ABOVE House designed by Pat Borer and Jon Broome for Glasgow Garden Festival 1988 and dismantled and re-erected in Fife to a different layout.

Adaptability for a long useful life

Buildings that can be adapted to changes of use – working at home, children, an elderly relative moving in and so on – have a huge role to play in minimising future emissions. Improvements in performance (better insulation and weatherproofing, for example), results in buildings that last longer and reduce the need for replacement with its attendant use of resources. We have considered adaptability as an element of a modern vernacular in Chapter 1.7. Converting unwanted deep-plan office buildings into much-needed housing could be a great way to make a profit, but it will never make good homes, as those offices were not designed to be used for anything else.

Durability

Along with adaptability, durability is another aspect of building for a long useful life. Commonly, buildings are designed to last 60 years and major refurbishments are often designed to last 30 years. However, commercial offices are often rebuilt after as little as 20 years and some residential buildings are so poorly planned and built that they are demolished well ahead of their 60-year planned design life and before the loans to finance their original construction have been paid off. In practice, most buildings have to last much longer than this. The Local Government Association suggested in 2017 that at current rates of construction and replacement, a house built in Britain today will have to last longer than the pyramids!

Fortunately, it is relatively easy to design buildings to last much longer than 60 years. As well as using parts that can be easily replaced when they are worn out or obsolete, designers are encouraged to use durable products and materials. This inevitably reduces the use of resources and energy to provide and maintain buildings over their lifetime.

Also useful is to specify materials that are self-finished and do not require applied finishes that need maintenance. Durable timber used in its natural state without an applied finish, brick, pebbledash or self-coloured render are examples.

Good maintenance is a very important part of reducing the use of resources and is a necessary complement to adequate specification of materials and designing details for durability. Designers should make sure there is adequate access above a glazed lean-to roof or to the ceiling of a double-height space and pay particular attention to services by providing plenty of access.

Tough, weatherproof materials should be used in exposed locations on the exterior of the building. Vulnerable details should be avoided, such as exposed parapets at roof level. Rather, details should be designed to protect the building, such as designing deep roof overhangs and setting windows back in the openings away from the face of the wall. This can have other benefits such as solar shading and rain protection if windows are left open for ventilation.

The durability of products and materials should be balanced against their environmental impacts. Some inherently durable materials require significant amounts of energy to produce and/or have a relatively high initial capital cost. That said, more valuable materials such as steel and aluminium will almost certainly be reused or recycled independent of any green policy simply because of their value, whereas composite plastics, for example, are more of a challenge.

Decisions for reduced upfront emissions

There is a hierarchy for reducing emissions. We all know the mantra 'reduce, reuse, recycle'. This can be expanded to the less catchy: 'avoidance, sufficiency, efficiency, specification, adaptability, durability, all with an eye on cost'.

The first two – avoidance and sufficiency – are the most impactful. They are at the same time the simplest and the hardest to implement. All we need to do is stop consuming stuff. We can do it now and it has a negative cost. If something is essential, such as food or shelter, then we can simply limit our consumption to what is enough. In our day job, we might realise that our client really doesn't need a new building or even an extension. Perhaps they just need to remodel the interior, so we could sell them that service instead. Perhaps they just need to get rid of their junk and don't need our services at all. How serious are we about sustainability? Do we need the money or can we reduce our overheads? It is carbon saving all the way down.

Other environmental impacts

Timber is seen as an alternative, low-carbon, renewable material to steel or concrete for structural purposes. However, sustainable timber has to be harvested from forests managed to regenerate and protect the environment. It also requires protection against

LEFT Whether to demolish or not is rarely a simple question.

decay and is flammable. It is less adaptable and weaker than both steel and concrete and therefore unsuitable for many applications, such as supporting heavy loads or spanning large distances.

Glulam, laminated veneer lumber and I-joists are stronger and more dimensionally stable than natural wood, permitting higher loads and wider spans. There is much excitement currently around the potential of cross-laminated timber (CLT – not to be confused with Community Land Trust). These laminated mass timber panels, 10cm thick or more, are suitable for walls, floors and roofs to construct multi-storey buildings (plyscrapers). The first tall timber building in the UK was nine floors constructed of CLT and completed in 2009; the highest completed at the time of writing, at 18 floors, was built in Norway in 2019, with a tower of 70 floors planned in Tokyo.

The supply of timber

The use of lightweight timber framing in low-rise buildings, together with its other uses in furniture, paper making and so on, is already exceeding the supply of timber from well-managed sources. A move to mass timber for high-rise buildings can only make the situation more critical. While the use of timber in construction may reduce (but not eliminate) carbon impacts on climate breakdown, it is having a detrimental effect on an emergency in biodiversity. Populations of insects are in steep decline worldwide, threatening agriculture, which relies on them for pollination, and the destruction of tropical rainforests is accelerating. The Living Planet Index managed by the Zoological Society of London (ZSL) reported in 2022 a staggering drop in biodiversity in the Caribbean and Latin American region of 94% over the last 50 years![5] Mixed woodland gives way to monocultural plantations of conifers with very low biodiversity, which are at risk from drought, pests and forest fires exacerbated by climate heating. The UK is one of the least-forested parts of Europe and there is plenty of space for more well-managed woodland, both for timber extraction and biodiversity enhancement.

Forestry, together with agriculture and urban space for human populations, increases the pressure on space on the planet. The overriding goal must be to reduce the quantity of raw materials required to build and maintain our buildings. Meanwhile, to paraphrase the life cycle expert Jane Anderson, 'We need more buildings in timber, not more timber in buildings'.[6]

Notes on the use of timber

Timber treatments are toxic to insects but also to humans and are to be avoided if possible. Durable or moderately durable wood does not require treatment if the timber elements are carefully detailed to avoid constantly moist conditions and if maintenance is good. The softwoods, Douglas fir and larch from Europe (including the UK) are described by Woodscape as 'moderately durable'.[7]

In new construction we can avoid details, which cause timber to get wet and prevent it drying properly. Timber will not decay if it gets wet, on the outside of a building for example, as long as it is able to dry out between rainstorms.

It should rarely be necessary to provide overall timber treatment in the UK. Most timber in a properly maintained building is not at risk and should not be treated. This is now reflected in the revised technical requirements of the National House Building Council and other building insurance schemes that no longer require interior timber to be treated. There is no reason at all to treat non-structural and decorative timber.

Plastics

On the topic of plastics, Charles Eames remarked: 'The problem with plastic is that you can do anything with it.'[8] Harmful substances are often added to plastics to create specific properties. Synthetics do not generally cause problems during use – rather after demolition. Dumped waste does not degrade and heavy metals may leach out. Incineration generates energy but can also create harmful emissions. Most plastics perform badly in fires and can release toxic fumes – wiring and windows are a particular concern. Many synthetics can be recycled but to do so, the recycled material must be uncontaminated.

Alternatives to plastics in construction

There are less harmful alternatives to most of the large quantity of plastics used in construction:

- Galvanised steel rainwater goods rather than PVC. Steel is also more robust.
- Linoleum, made from natural raw materials; linseed oil, pine rosin, wood, cork and jute, rather than vinyl floor tiles and sheeting. Lino costs more but is longer lasting.
- Polyethylene (PE) for above and below ground drainage instead of polyvinyl chloride (PVC).
- Timber windows and cladding rather than PVC. PVC windows do not require decorating and are commonly specified to reduce maintenance, however, they are not necessarily a zero maintenance component as they can become brittle with age and suffer from corrosion of the steel core of the window sections. Proper detailing of timber windows can provide a long-lasting solution (see Chapter 3.5).
- Low smoke zero halogen cables are an alternative to conventional PVC-insulated cable. They incorporate a high level of inert filler into a thermoplastic insulating material. They are safe in a fire, which is particularly important in large public buildings such as theatres.

This is an expanding field both in terms of available products and also in terms of regulations and the science and evaluation of materials. Material and product choice is also dependent on the scale and type of building and where it is in the world.

The more details, the more devils

The way buildings are put together has been devised over the years to achieve weatherproof, comfortable and safe environments, which are warm, dry and safe against collapse and fire. The need to design and build for a future that will address the need to conserve fuel and materials and counter the effects of extreme weather creates new requirements and problems to solve; principally, ensuring the building envelope is:

- weathertight against severe rain and wind;
- airtight;
- free from thermal bridges;
- minimises environmental impacts
- designed to last a long time and not require a great deal of maintenance and repair.

There is, as usual, no one way to design the envelope of a building, but some basic principles apply which we tend to follow to create an economic, relatively straightforward and risk-free solution. As with all design, detailing can be approached as problem-solving or an exercise in visual style.

Detailing the building envelope

A good starting point is to consider the building envelope as layers that ideally don't cross over each other. If we consider a wall, we might have the internal finish, then structure, insulation and finally a windproof and weatherproof external skin. We need a continuous airtight layer to prevent unwanted infiltration, which causes cold draughts, energy loss and risk of moisture damage to the envelope. Generally, this needs to be on the warm side of the insulation in a predominantly heating climate. The most basic solution would be to use the internal finish, for example wet plaster. This can work but we also need to accommodate the services, such as wiring, that are likely to run in external walls. We could install wiring and boxes into chasing in a masonry wall, but the wiring will create air paths and will stop the plaster sealing the porous masonry unless we pre-seal the chasing before installing the electrics.

If we use a dry form of construction such as timber frame then we cannot rely on the plasterboard as the airtight layer. Even if we could seal all the electrical sockets and other services penetrations, we would struggle to achieve continuity of airtightness behind floors and internal walls. It is far more effective to use a service void on the indoors side of a separate robust air barrier; see the schematic below.

The wind barrier is a crucial component to reduce thermal bypass and it also provides a second line of defence against rain if a separate rain screen is employed. This means that, as with the air barrier, we should design out penetrations as best we can and take care when detailing any unavoidable ones.

Ideally, the insulation layer should also be continuous with no gaps, including between it and the wind and air barriers. Sometimes insulation is fitted between structural elements such as timber studs - this can be fine as long as the timber fraction is low and the insulation is a tight fit. In practice this tends to mean using 'I' studs and a densely packed insulation such as mineral wool, blown cellulose or recycled paper.

The amount of insulation and the optimum position and vapour permeability of the air and wind barriers will vary with climate, but the principles of maintaining continuity are universal. For predominantly heating climates, such as the UK, a good rule of thumb is to

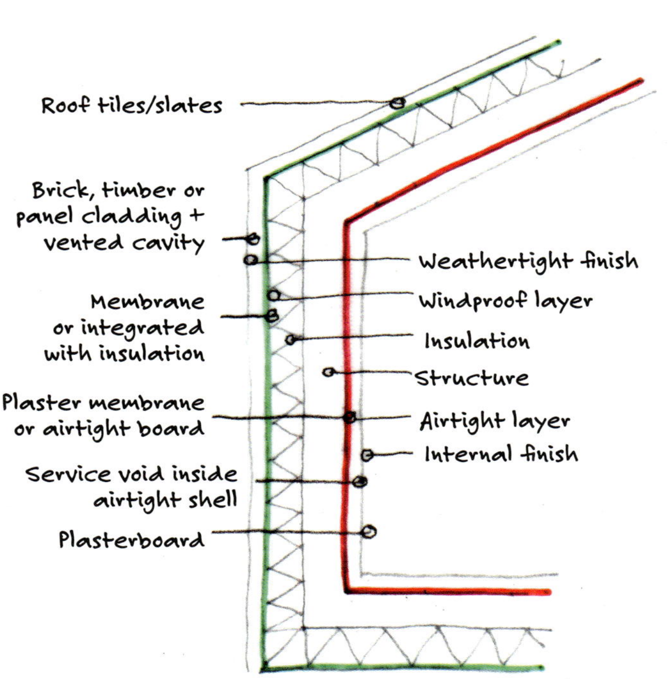

LAYERS IN THE EXTERNAL ENVELOPE

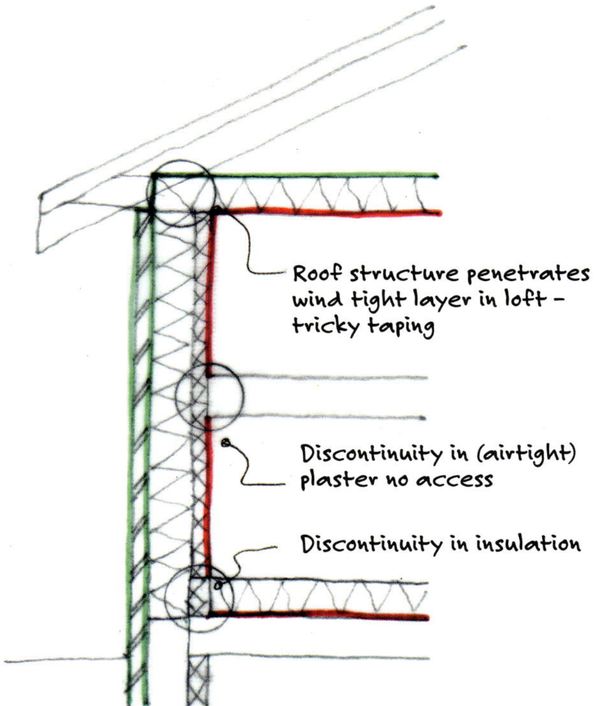

TYPICAL ISSUES WHEN LAYERS CROSS

ensure that the external wind barrier has at least five times the vapour permeability of the internal airtight layer. Both layers should be airtight, although we only usually test the airtightness layer.

This generic diagram could cover a range of construction methods. The structural layer could be precast concrete or CLT but if it was timber frame then it would most likely include some or all of the insulation, possibly with additional insulation over to reduce thermal bridging if the timber is solid studs rather than 'I' studs.

If the structure is steel frame then we will need to insulate on the cold side of the steel, although there may be some insulation between the studs as well, largely for acoustic reasons. If we put the air barrier on the warm side of the frame, we need to achieve continuity where floors and internal walls join the external structure. As most of the insulation will be on the cold side of the frame it makes sense to simplify things and place our air barrier on the outside of the structure followed by insulation.

Generally, we want to avoid anything going through the airtight, insulation and weather-tight layers as these will be hard to seal, adding cost and risk. However, in this example the roof structure penetrates the insulation layer, forming a thermal bridge. This can be ameliorated by employing a thermally broken fixing. Sometimes the roof structure

ABOVE Porotherm clay. Light clay blocks with thin mortar jointing promise a simple monolithic construction and have been used for Passivhaus buildings in the UK climate.

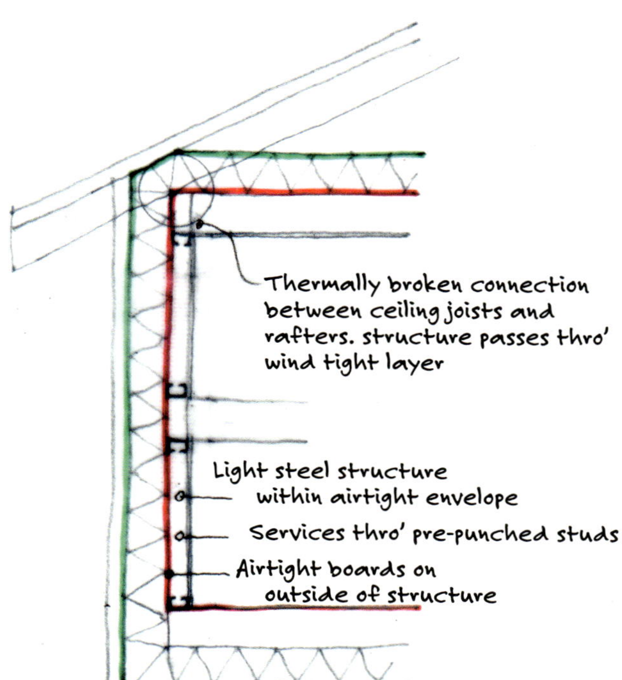

STEEL STRUCTURE WITHIN AIRTIGHT LAYER

is a conventional timber trussed rafter on top of the steel structure, which reduces the thermal bridge.

Our external weatherproof layer might be heavy brick, stone, lightweight timber, composite panels or metal. All will need some level of connection back to the structure and, again, this can compromise the insulation as a more conductive thermal bridge or by creating gaps in the insulation that allow convection within the structure.

The apparent exception to the layering rule is monolithic construction. Solid stone or brick buildings are monolithic but lack insulation. The historic exception is structural straw bale building, where the bales provide insulation and structure. Another example is light extruded clay blocks. These are a finer version of a traditional extruded clay block but with much lower conductivity thanks to the fine honeycomb section, which may even be filled with insulation. These blocks can be rendered on the outside and plastered on the inside. The conductivity is higher than for normal insulation, but the thicker blocks have been used to build Passivhaus homes with a good form factor. We need to think about how to deal with the services by chasing and parging, creating a service void after parging, or either avoiding wiring in the outside walls or surface mounting it.

We have considered a wall, but similar principles apply to floors and roof. Generally it is easier if the roof and floor layers are in the same order as the wall, otherwise we will have to think about how to join them up.

The envelope details need consideration at an early stage. By the planning stage, many problems will have been either avoided, solved or created. Even if the method of construction is to be decided later, it is crucial to think about how the functional layers of the building will go together.

Aesthetic detailing

When we say a building is 'nicely detailed' we are often talking about the finishes that we can see rather than how the vapour control layers in the roof connects neatly to that of the walls and floor. Weather resistance and good thermal performance won't win you any architectural awards, but will hopefully provide a deeper sense of satisfaction, while also keeping you out of court.

Aesthetic details are far more subjective and while we have our own stylistic preferences and pet hates, we try hard not to judge on looks. One of our themes in this book is that in general, style should not undermine function or affordability. Clearly, most of us regularly embrace reduced function and higher cost in the name of style but there has to be a threshold. Nick rides vintage motorbikes wearing a wax cotton jacket and Jon designed a - sometimes rather chilly - fully glazed, double-height living room for a view of the trees in the garden. The problem is when someone else is paying a potentially high price for their designer's personal stylistic preferences. We can take off an uncomfortable shoe after the party but we have to live and work in our buildings.

Perhaps as a reaction to what are seen as traditional clichés, new clichés have emerged. The traditional skirting board and architrave have given way to shadow gaps. Roof overhangs, barge boards and gutters have been clipped and concealed. Trim, such as skirting boards and architraves, serve a number of really useful functions. They protect vulnerable areas from damage and greatly simplify construction. Anyone who believes that eliminating them will simplify construction and save money has never spoken to a plasterer.

By all means, come up with a better solution to resolve the connection between two different elements of construction but don't just adopt a more expensive and less practical cliché and think it is sophisticated.

Juraj Mikurčik's self-build home (see pages 108 and 150) is a very simple cost-effective design with all the envelope layers nicely resolved. He is an architect and this is his own home so he decided to indulge in some minimalist details at his own expense. For example, the clay plaster finishes flush with the internal door frames. It looks lovely but only another architect would notice. If there had not been the time or budget to do this, then a simple timber trim would be a different solution and the home would be no less lovely.

The roof is cheap black 'corrugated iron' with a good overhang. Juraj decided to leave off the gutters and see how it worked. He was prepared to add a short length by the entrance if the runoff was a problem when going in and out of the house during rain, or for the whole roof if the idea didn't work out. In practice it has been fine and is a nice example of genuine minimal design or radical simplicity.

Some useful general principles for detailing

Jon found himself early in his career using drawings to communicate how Walter Segal's houses are assembled to self-builders unfamiliar with building and unused to reading technical drawings. He learned ways to facilitate this process, which have guided his practice since. Useful practical principles include:

- Imagine you are carrying out the assembly onsite yourself. Establish the sequence of operations one step at a time - measure, cut, screw, seal, finish, *etc.*
- Keep details as simple as possible. This saves time and money and reduces the risk of details not being executed properly.
- Identify difficult details - such as thermal bridging at structural overhangs and balconies, weathering at parapets, airtightness around service penetrations - and design them out as far as possible.
- Do not express the structure on the outside - roof trusses above the roof, columns outside the walls - and avoid cold, draughty roof voids.
- Detailed drawings need to explain how things are fixed together, with adhesive or screws or both; where clearance is required or where a tight fit is needed; where a weatherproof or windproof seal is necessary. Consider realistic dimensional and angular tolerances for factory- and site-made components.
- Only show the elements and information that the builder needs to execute the task in hand. Avoid extraneous information that may confuse and hinder understanding; only show dimensions necessary to position the elements in question - not more.

- Relate the assembly in question to a grid line and/or a part of the building that will be in place at that stage of the construction. When doing this, make sure you are not referencing something that will not be present until later in the building process.
- In this connection, decide at the outset whether you will be dimensioning to main structural elements, which will be in place early on in the construction process, or to the face of finished elements, walls and floors, which will not have finishes until later in the process (*e.g.* plasterboard).
- While on the subject of dimensions, digital drawings have advantages over hand drawn ones as they can be easily amended, for example, though they can give a misleading impression about the accuracy necessary and achievable in construction. Millimetres are a very small measure to use on a building site where it is equal to the thickness of a pencil line. Hand drawings with their slightly wonky lines, on the other hand, give a visual signal of an appropriate level of accuracy.
- Cut away 3D drawings are a very good means of showing how the parts in a wall, a window and a floor, for instance, fit together as they can combine three elevational and sectional views in one image.

In addition, Walter Segal rethought construction from first principles as well as fully integrating structural design and cost control into his design process. The startling simplicity and economy that followed this innovative and integrated approach has informed the work of some designers to this day. Although Segal buildings cannot meet current standards of insulation and airtightness, there are some lessons with general relevance:

- 'Dry' construction eliminating all 'wet' trades: concrete, brickwork, plaster. This makes construction accessible to people without building skills and the building can be easily altered, improved and extended.
- Although exposed timber gets wet, it does not rot because it is fully open and does not remain wet for a protracted period, which would otherwise lead to decay. This strategy informs an approach to timber detailing generally, with end grain always left open to the air and not butted tight so that air can circulate freely.
- Segal designed 600mm-deep roof overhangs which protect the building effectively against the weather and reduce the need for maintenance.
- He devised ways of making stairs both internal and external without having to make complicated angle cuts, relying instead on screwing together hangars, posts and treads at right angles to one another.

PROUD TO
BE A 5 STAR
BUILDER
Sales C

Elements of good building

There are a large number of possible ways of building, but in the context of this book we are interested in examples of problem-solving to make buildings better.

For most buildings, foundations and ground-floor construction is where a lot of the cost, upfront emissions and risk lies. Our experience is limited to low- to mid-rise buildings, but it is these smaller buildings where the cost and performance can be dominated by foundations. Materials in contact with the ground need to resist moisture. Load needs to be transferred to the ground but we want continuous insulation and wind tightness to reduce unwanted heat loss due to thermal bridging and thermal bypass. We might want the floor to have a bit of spring rather than being hard and unyielding or we might need it to take load from internal walls or heavy equipment free of vibration, such as in a medical diagnostic centre.

The ground might be stable and load bearing or it could be heavy clay with nearby trees and subject to shrinkage in dry weather. It could be a green field or made-up ground of variable depth and we might have to deal with natural radon gas or unnatural contamination from the site's previous use. There can be a lot to consider and it should all influence our choice of construction.

Outside the UK, basements are much more common. These tend to be concrete, will often need to be fully waterproofed and are very carbon intensive. We have mostly managed to avoid building them but it is ironic that our main experience involved basements for biomass boilers and fuel stores in education sites.

BELOW Alternative foundation and structural combinations for timber construction: post-and-beam frame on self-buildable concrete pads, timber platform framing on concrete piles for very poor or very steep ground and finally steel screw piles for speed.

Sitting lightly on the ground

Jon has a lot of experience of self-build projects with small budgets, and many of these have used post-and-beam construction resting on very simple pad foundations.

Groundwork is minimal and can be done without machinery if access is a problem, and concrete can be avoided. A raised ground floor makes more sense on a sloping site where the building can 'float', creating useful storage space underneath. The ground can remain in its natural state without disruptive earth moving to create a level 'shelf' for construction. This is particularly relevant for those building on steep slopes or in woods, bogs or on the beach. As the floor is off the ground and ventilated, it can be made from timber with organic insulation further reducing carbon emissions. Indeed, it is possible to completely eliminate concrete by using stone pads, crushed rock or steel screw piles. Where DIY enthusiasts have skimped on excavation leading to initial settlement, it is a simple matter to jack up a corner of the building and add some shims!

There are disadvantages to creating a void under the floor because it increases heat loss. It also creates problems providing level door thresholds, and access to the void is difficult on a level or only gently sloping site. If we can lose spoil on site then levelling a sloping site may not be so expensive and will create a more usable plot. For many projects, this is the only practical option.

Segal homes use post-and-beam frames which penetrate the floor. This makes it much harder to get the floors truly wind- and airtight. Others have constructed an insulated timber platform and then built load-bearing timber walls off this to create a highly insulated, airtight construction. The illustrations below show these options.

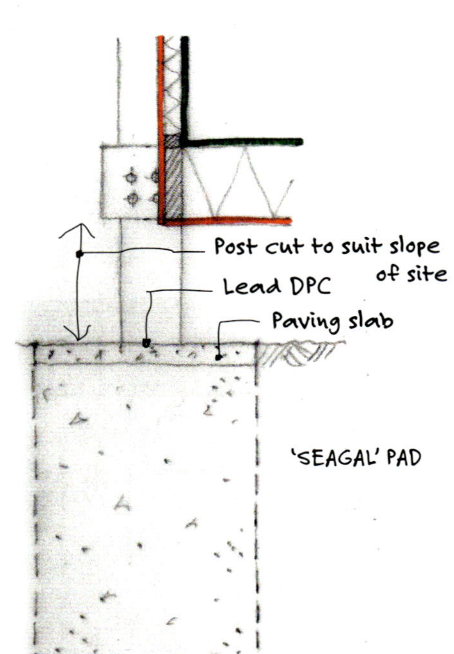

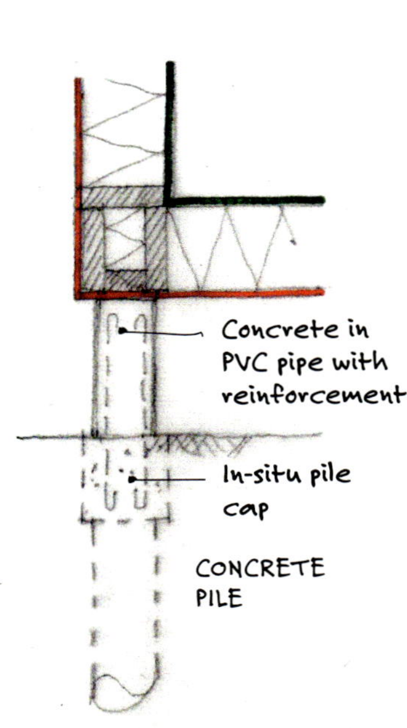

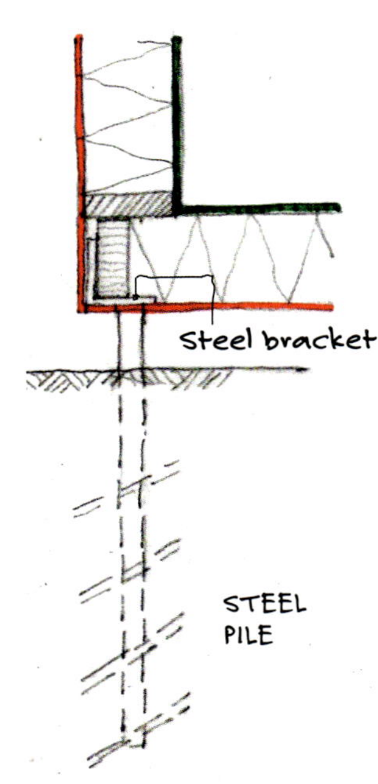

BUILDING ABOVE THE GROUND

Building on grade

Building on grade reduces the heat loss through the ground by around 40% for even a small building. Larger ground-bearing buildings may need little or no insulation under the floor as the ratio of floor area to perimeter is greater and the soil does most of the insulating. Soil is not a great insulator but if the heat path is long enough, this does not matter so much. We can, of course, make a suspended floor 40% thicker and fill it with cheap organic insulation rather than the more expensive load-bearing insulation we would need under a ground-bearing floor. To do this, we need to ensure the underside is completely windtight and vermin proof.

Most buildings in the UK are built on the ground. Conventionally, a concrete slab on the ground is founded on concrete block walls on concrete strip footings up to 900mm deep, or concrete trenchfill if it is required to go deeper – when close to existing trees, for example. Insulation is on top of the slab to provide continuity and avoid a thermal bridge. It is often required to raise a timber frame above the slab level to reduce the height of the ground floor above the external level. This necessitates a construction resistant to moisture between the slab and the timber frame. The example shown below is an autoclaved aerated concrete (AAC) block, however this can be awkward to construct and is a weak spot in the insulation layer as it transmits heat three times more than common insulation materials. We also worry about any water from plumbing leaks, spills or rain getting trapped between the insulation and the slab during construction.

Constructing the building on a concrete raft slab on a layer of consolidated selected fill spreads the loads and avoids the need for deep concrete foundations.

ABOVE Insulation below slab. Proprietary edge insulation units create formwork to the edge of the slab which will be laid on top of insulation panels.

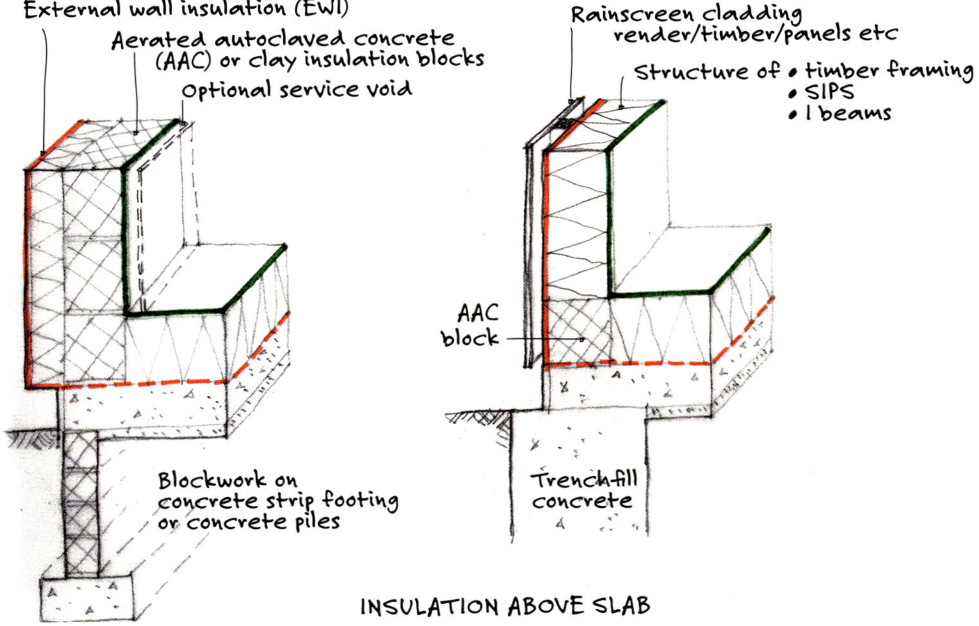

LEFT Foundations of concrete blocks or in-situ concrete trenchfill with concrete or clay block structure and either external wall insulation (EWI) with render finish or lightweight timber studs, I beams or Structural Insulated Panels (SIP) with timber, render or panel rainscreen finish. Option of service void.

If we put the insulation under the slab, we can solve the issues of moisture risk and thermal bridges. Nick first heard about this approach from the energy consultant David Olivier and adopted it for his self-build home. The main advantages were relative simplicity of construction and continuous thermal-bridge-free insulation. Concrete rafts are quite common for industrial buildings but in 1995 it was considered unusual to put them on top of insulation. The engineer embraced the challenge, but his first intuition was to thicken the slab edge like a traditional raft. This would greatly complicate the insulation, damp-proof membrane and steel work, so a flat slab was designed instead. DIY or commercial versions of this approach are now the default solution for small- to medium-size low-energy buildings in the UK, from homes to schools.

Although the slab is structural, the total amount of concrete can be less than for a conventional concrete ground floor with strip footings. The slab can be ground as a floor finish and underfloor heating pipes can be fixed to the bottom layer of mesh.

This approach can be used with masonry construction, but in our timber example it raises the timber above ground level for protection. The same approach was used for some of the first Passivhaus schools in the UK but with the slab level with the surrounding ground to allow access to outside from all the classrooms. A simpler solution to the AAC upstand described above was adopted with the timber structure at slab/ground level, but with the EPS or XPS edge insulation continued to about 200mm above ground. This insulation can be rendered or brick clad, with the insulated timber zone starting above this. Although the timber structure is at ground level, it is warm and dry and at no more risk than the furniture and internal walls.

INSULATION BELOW SLAB

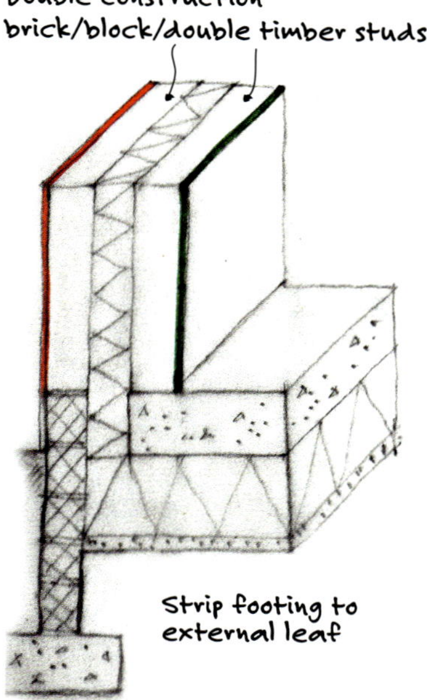

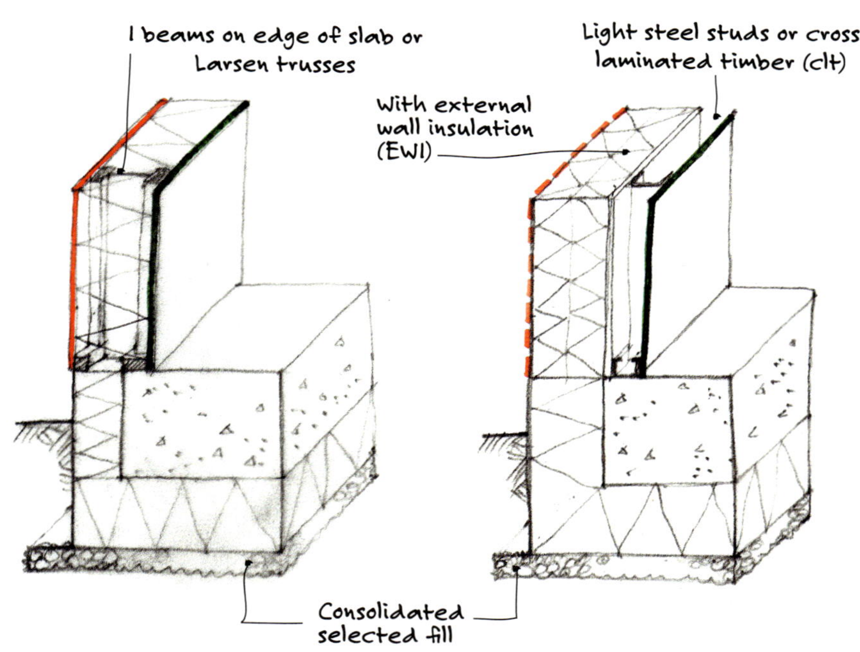

For larger buildings, piling might be required, reducing the practicality and effectiveness of insulating under the slab. A simpler solution is to use a combination of vertical perimeter insulation and thinner insulation (if it is actually required) on top of the slab.

A larger multistorey building will only lose a small proportion of heat through the ground floor and the thinner insulation required reduces the risk of any moisture problems.

Concrete-free ground-bearing floors are possible but are more common in North America. Architects Polly Upton and Kirk Rushby built their Passivhaus on a site with no access for a concrete lorry. The access constraint gave them the incentive to try a less-conventional solution. The walls rest on foam glass blocks and the floor comprises two layers of plywood resting on 200mm of EPS insulation. A membrane forms an airtight flexible connection between floor and walls.

Their structural engineer, Andrew Collinson, went on to build his own Passivhaus using a similar approach but with cheaper AAC blocks to support the walls. He also wanted to avoid using oil-based insulation so used foam glass gravel under the floor and walls with a limecrete and earth floor on top. Foam glass gravel probably has higher upfront carbon emissions than EPS but like the limecrete and earth floor, the choice was more aesthetic.

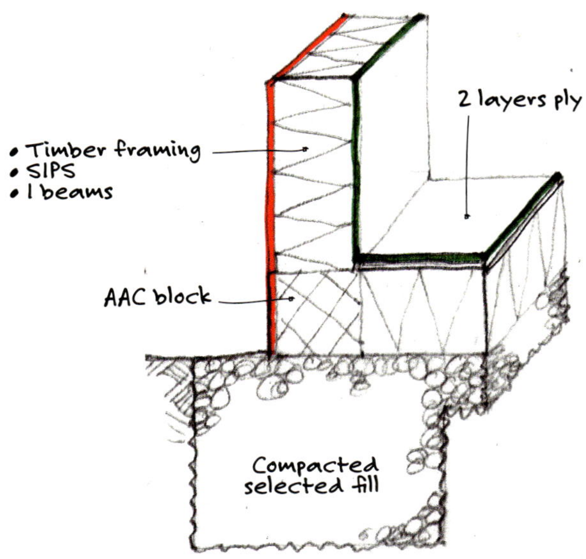

ABOVE Foam glass blocks under load-bearing walls EPS slabs will then be laid on compacted selected fill for Polly Upton and Kirk Rushby's concrete-free foundations.

LEFT Concrete-free foundation with a floor finish of two layers of plywood and foam glass blocks below load-bearing walls.

Walls

Walls can be load-bearing or an infill or cladding for a structural frame of timber, concrete or steel. Frame structures have the advantage over load-bearing walls of being more adaptable. The non-load-bearing infill can be moved and upgraded to suit changing uses and performance requirements without affecting the structure. For houses and medium-rise buildings, load-bearing walls are more economical as they avoid doubling up on structure.

In the previous chapter on detailing, we see how the layering of the various components is an important consideration when designing a wall. There are a multitude of potential solutions that will work well, but there are even more possible options that will perform poorly in in terms of cost, durability, energy efficiency and upfront carbon emissions. We will assume that fire and structural requirements will be met, although we know that is not always the case.

For non-load bearing walls, the main challenge is how they connect to the structure. The main options are as infill, as a continuous 'wrap' on the outside of the frame, or a combination of the two. External structure has been a fashion, but we hope the lessons have been learnt.

Infill in the traditional sense of historic timber-frame buildings is no longer an option as it cannot meet building regulations U-values, even on paper. More usual is to infill a concrete or steel frame with blockwork, light steel frame or timber panels to form a flush external surface that can carry a continuous layer of insulation. If the frame is timber or steel then it will be most effective if the airtight layer is on the outside of the structure, but with a concrete frame the airtightness could be on the inside of the panels, with each panel sealed to the airtight concrete frame and floors.

If the walls are to wrap the frame then the challenge is how to make them airtight and how to fix them back to the frame. Where airtightness is a challenge, we can assume acoustic, fire and smoke separation will also be harder to achieve reliably.

MULTI-STOREY FRAME

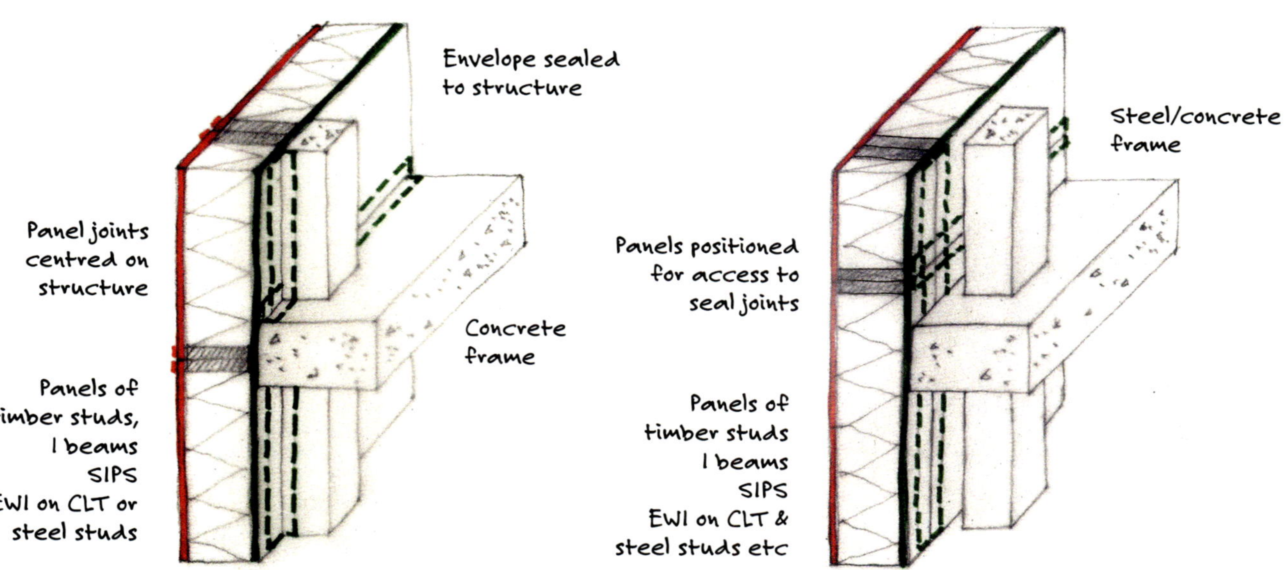

Cladding

The external cladding is the next challenge as this will need to be fixed back to the primary structure without creating significant thermal bridges. Self-supporting brick façades may only need simple ties, but where masonry support systems are used these will incur significant thermal bridging and potential for thermal bypass.

Lightweight ventilated façades are typically fixed with adjustable helping hand brackets supporting profiles to which cladding materials are fixed. These usually form a significant thermal bridge, although manufacturers often do not provide values. They are also difficult to install well and introduce a risk of thermal bypass by creating gaps. Although properly thermally broken brackets are available, these may not meet requirements for wall components to be non-combustible. Hopefully by the time you are reading this, the market is flooded with low-cost, thermally broken, non-combustible façade support brackets, but there is a simpler solution. Faced with this problem, RDH Building Science consultants in Vancouver experimented using high compressive strength mineral wool insulation and vertical timber or steel battens fixed through it with long stainless-steel screws. The screws alone would simply bend, but the dense insulation takes significant load and the resulting assembly is very strong. This is another nice example of practical experiment and value engineering as the solution developed is cheaper, easier to build and has improved thermal performance.

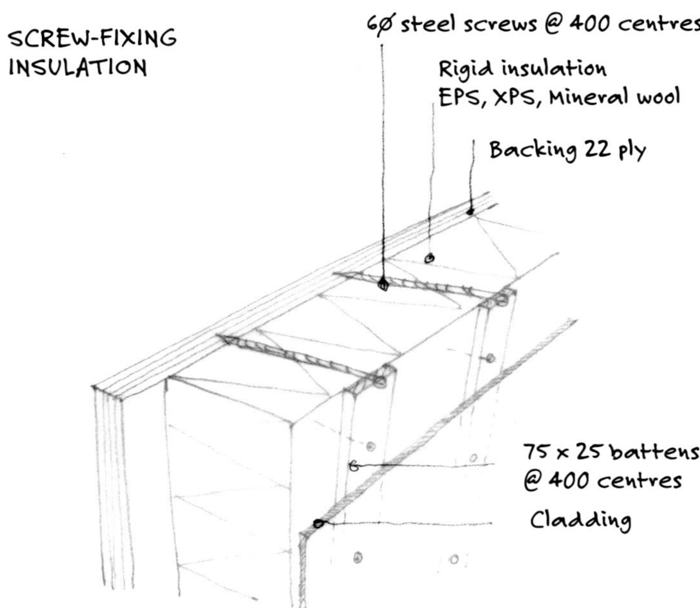

 A simple and efficient alternative to façade brackets to fix cladding.

Frames for residential buildings

The Spreefeld community development in Berlin comprises three concrete-frame buildings with a timber-framed, airtight, highly insulated and triple-glazed external wall wrapped around the structure. The project shows the inherent flexibility of a frame, accommodating as it does single- and double-height spaces housing conventional apartments as well as multi-generational dwellings housing up to 21 people as shown in the plans which also show the fixed structure and service ducts in black. There are also shops, cafes, kindergarten, workshops and other uses within the buildings. Each of the three buildings is organised as a co-operative with sub-groups all under a single overarching co-op.

Load-bearing walls are generally easier to detail. They may be site-built from timber or masonry or prefabricated as panels in a factory. There has been renewed interest in double-height panels (balloon framing) recently because it has the potential to make

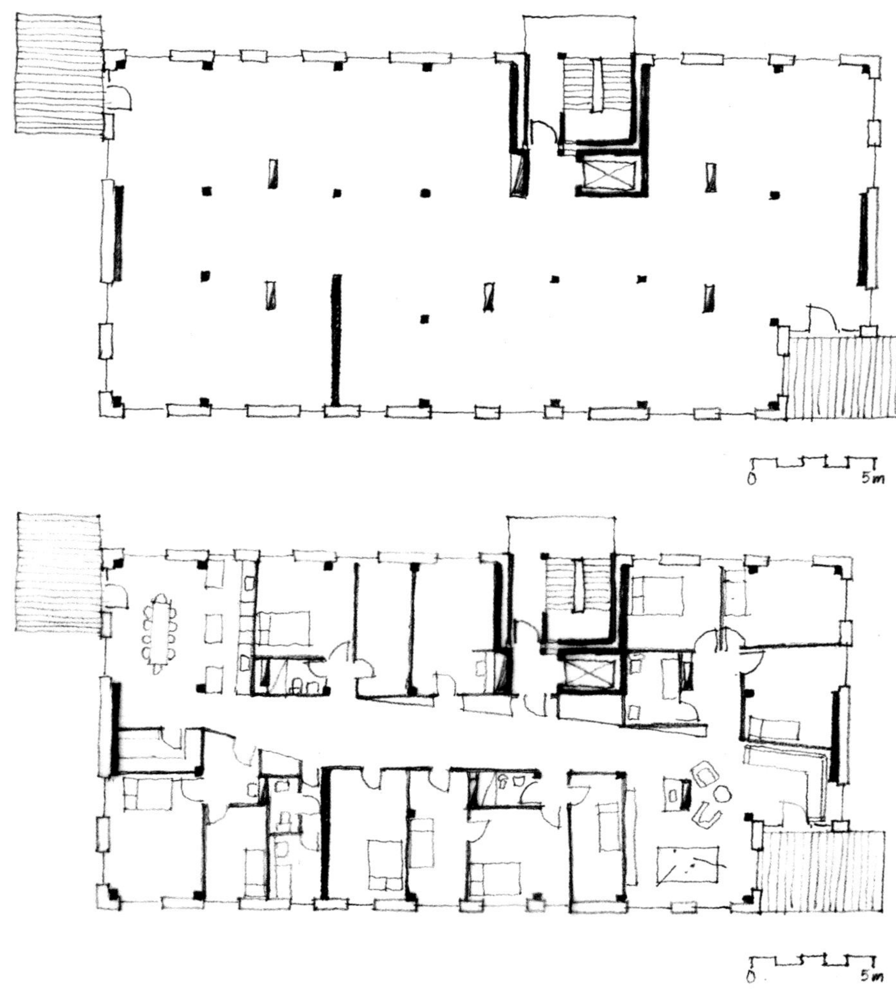

TOP Spreefeld showing the three buildings that make up the complex with a kindergarten in one of the social/commercial spaces on the ground floor.

BOTTOM Spreefeld Interior showing freestanding column in the corner of the room. The well insulated envelope of the building is freestanding outside the line of the structural frame.

achieving an airtight envelope in timber much easier; in platform framing the intermediate floor penetrates any airtight layer on the inside of the external wall. This forms a thermal bridge and an inherent weakness in the system, which can be hard to make robustly airtight. In balloon framing, the floor is supported on a ledger fixed to the face of the wall panel without interrupting the airtight layer.

The same principle can be applied to light steel frames or even masonry construction, with floors supported on a ledger rather than built into the wall as is traditional. For off-site construction, some manufacturers are using full-height panels with vertical joints for this reason. Others split the walls horizontally but continue the ground-floor wall panels above the first-floor level. The floor is supported on a ledger and the second storey is planted on top of the first with the airtight seal easily accessible.

Potential materials

Most of the above principles can be applied to different materials, but materials also have their own constraints and benefits. The following list of potential materials is not exhaustive but is intended to highlight the influence of material properties. Our personal preference is to build in timber but that is not always possible. The case can be made for the fire resistance of timber construction, but if insurers insist on wall assemblies that contain no combustible materials then other solutions may have to be found.

INTERMEDIATE FLOOR

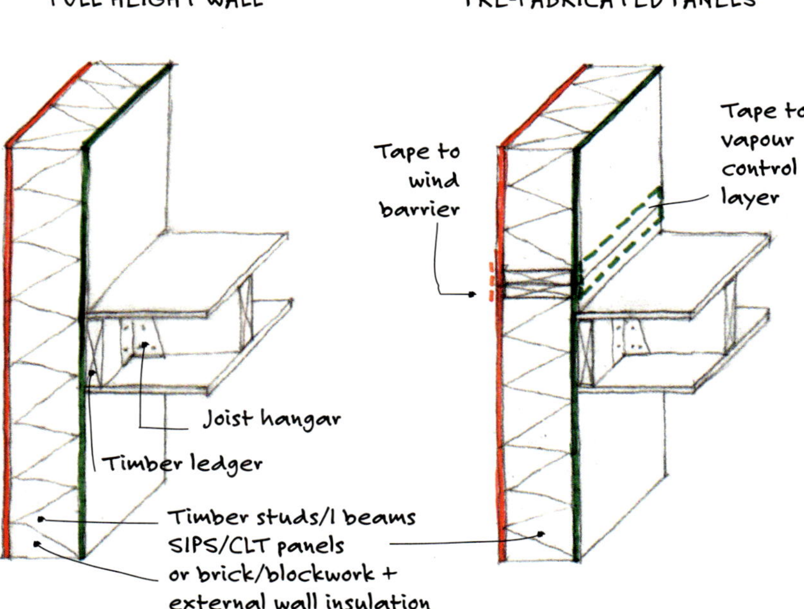

Timber 'I' beams

'I' beams reduce cold bridging to a minimum and can create a very simple, robust, well-insulated, airtight envelope.

To avoid thermal bypass, the insulation needs to fit the non-rectangular void between studs. Dense blown fibre, such as recycled newsprint, is the favoured solution but layered batts of fibre-based insulation can also be used with care.

Cross-laminated timber

Panels of solid laminated timber for floors, walls and roofs can be used to construct high-rise buildings. This simple construction with screwed connections has a number of advantages. It has good soundproofing and fire resistance characteristics due to the weight and thickness of solid timber and the ability to fix anything anywhere, from light fittings to lift guides. It can also be shown internally if treated for surface spread of flame. However, the amount of timber involved is at least three times as much as required for conventional stud framing and would endanger timber supplies if widely adopted (see Chapter 3.2).

Concrete

Residential concrete-frame apartment buildings are commonly designed on the continent with a highly insulated external timber-frame envelope fixed to the outside of the frame to eliminate thermal bridges. They have blockwork fireproof and sound proof internal partitions, which can be demolished and rebuilt in a different arrangement, if required in the future.

Insulating blocks

Load-bearing blocks with low conductivity can be used with a thin-joint mortar system to produce a simple monolithic wall. One example is AAC, which is usually used as an inner leaf for cavity construction. Another option is light extruded clay blocks. The wider versions can be used for Passivhaus construction, either on their own or with additional external insulation. Wind- and air-tightness can be formed by external render and wet plaster respectively. Services can be chased in with care to maintain a continuous airtight barrier or a service void can be created for wiring in external walls.

Steel

Steel frames using heavy hot-rolled sections are widely used in commercial construction but smaller-scale buildings generally use lighter cold-rolled sections, which can be drilled and screwed together using handheld power tools to form load-bearing panels. Steel studs, which are dimensionally more precise and stable than timber, are also used for internal partitions.

Despite the advantages, steel is more challenging for airtightness, thermal-bridge-free construction, acoustics and fire protection. At the time of writing there are very few examples of Passivhaus buildings in steel. The rapid uptake of Passivhaus in Scotland is leading to progress and contractors and designers working together to find viable solutions.

TOP Hanging floors off the walls keeps the layers aligned and simplifies airtight details. Inspired by timber balloon framing, the principle can be applied to CLT panels and light steel construction.

ABOVE Precut I beams and notched sole plates speed up on-site stick building at lower cost than offsite panel construction.

Roofs

For most housing and smaller buildings, pitched roofs are the safest option as they shed water by gravity and are easily vented. There is a huge choice of materials for the primary weatherproof layer to suit all styles and budgets.

The form of a pitched roof can limit the plan arrangement of the building, although this constraint will improve the form factor and reduce costs. Future extensions can always adopt a flat roof, if required. Pitched roofs often have trussed rafters and insulation at ceiling level. Traditionally, the cold space above is well ventilated to avoid moisture issues due to air from the building leaking into the cold loft space. This can lead to thermal bypass due to wind-washing, especially at the eaves. This can be solved by avoiding ventilation but using a windtight but vapour open board or membrane over the rafters that joins to the wall's windtight layer. The ceiling must be airtight to reduce the flow of warm, moist air entering the roof space, and access is best limited to inspection and maintenance, rather than for storage.

Many builders of high-performance buildings are tending to use a cathedral ceiling with insulation between timber I-joists run from a ridge beam to eaves or from gable wall to gable wall as purlins. The warm side has an airtight board and the cold side has a breathable board and membrane with a ventilated and drained cavity created by battens and counter battens. The heat loss area is increased, but interesting and useful spaces are created under the roof. Depending on the pitch and ridge height, mezzanine sleeping or study galleries can be introduced as well as dry storage within the thermal envelope. Conventional wisdom says that trussed rafters and flat ceilings are cheaper but some designers and builders only use cathedral ceilings, arguing that they are actually cheaper and easier to do well.

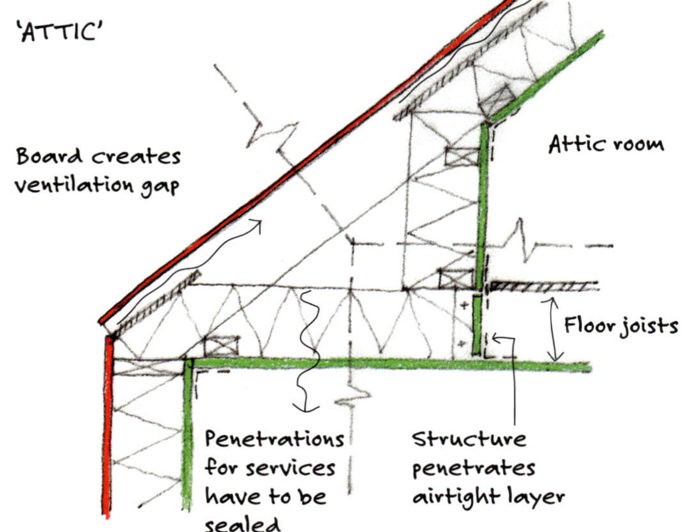

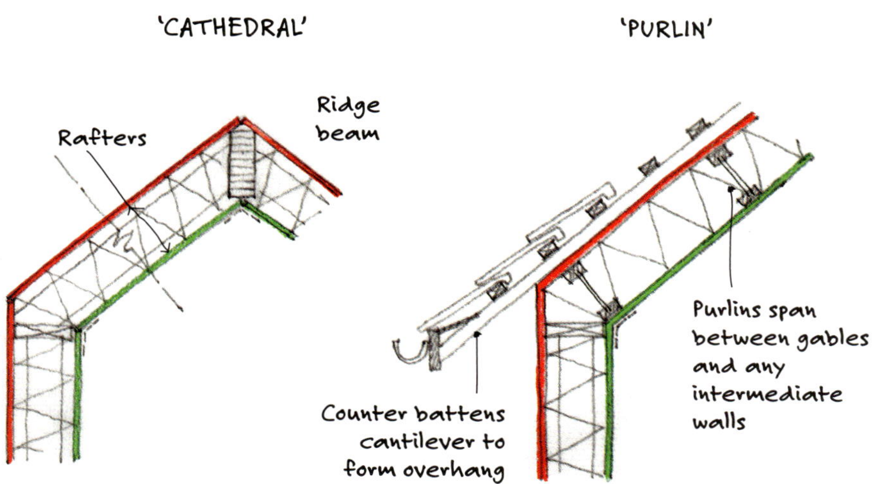

Larger buildings will often have flat roofs, although shallow pitched roofs with metal cladding can achieve considerable spans. Flat roofs have a bad reputation for failures caused by leaks and condensation. Pitched roofs rely on their slope to shed water quickly, which is why we can use almost anything to create a reliable rain barrier, from straw to flattened oil cans. Nowadays, a secondary vapour permeable membrane is used under this layer with a drained and vented zone between, providing a robust weather barrier.

A flat roof, by contrast, will need to be completely watertight as rainwater will pool. Being flat, it will almost certainly be used for maintenance access and may be used to locate plant such as chillers and air handling units. This makes it more prone to damage. Being impervious also means that water vapour cannot escape. Smaller flat roofs can include a ventilated zone, like a flatter version of a cathedral ceiling, but for larger roofs this becomes impractical. Parapets reduce the potential from cross-ventilation and require upstand vents, which need to avoid water ingress.

One very robust solution is to place the waterproof layer under water-resistant insulation, such as XPS, which is held in place with ballast. One issue here is that water draining through the insulation will significantly degrade the thermal performance. The relevant standards provide correction factors, but these will never be accurate and – like corrections for gaps in wall insulation – are rarely applied.

Finally, even if you have the perfect system, you have to consider what happens if it rains during installation. A partially completed flat roof will collect a considerable volume of water, which will overflow at the lowest points. These will not yet have the gutters and hoppers in place. This is a particular concern for our favourite construction material, timber, which will need to be fully dried out to avoid rot.

All that said, Jon has specified loose-laid roof membranes rather than fully bonded ones for many years without problems. We have to consider roof access by contractors

ABOVE I beams as purlins eliminates the need for ridge beams or ties and allows windows close to ceiling level without adding lintels in the roof insulation plane as there are no rafter loads but it may require a structural internal partition across the building midway between the end gable walls.

who may or may not take enough care. Ballasting the roof protects the membrane from physical damage and the effects of ultraviolet radiation, which lengthens its lifespan. A green roof buffers storm water from surcharging the drainage system, modifies the microclimate in towns and also creates habitats for wildlife. However, you will have to increase the load-carrying capacity of the roof structure to carry the additional weight.

Roof overhangs

We are big fans of roof overhangs and traditional – rather than so-called 'hidden' – gutters. Aesthetically this is a matter of opinion, but our preferred look also provides useful shelter from rain and sun and is easy to ventilate, even if flat.

Sometimes parapets must be used, which have their own challenges. Products are available to help with the thermal bridge created if building in concrete. Light steel frame is more challenging. Typically, the airtight line is outside the wall structure but at ceiling level for the roof. It seems to be difficult to connect the two layers through the parapet and also to create a thermal break whilst maintaining structural integrity. There will be a solution, but we don't know what it is yet.

For timber – whether CLT or stick built – we have good details to avoid thermal bridges and achieve a continuous airtight line. Our concern with timber is water leaks or rain entering the walls and roof structure before the roof and outlets are complete.

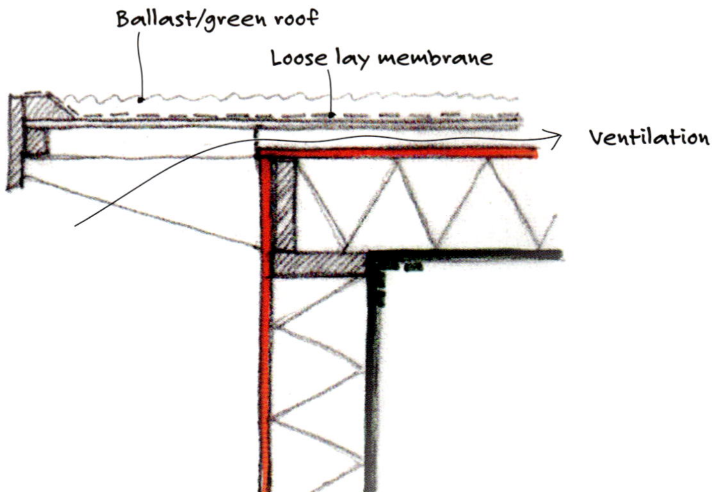

The anatomy of a window design

The domestic window has evolved over time, driven largely by function and constraints. We see variations between countries with different climates. The following examples are typical of the UK.

From the mid-eighteenth century, the classic Georgian sash became popular. It is tall for good daylight penetration and allows reasonably controlled ventilation from one side of a room by opening the top and bottom. The panes are small because float glass, formed by floating molten glass on a bed of molten tin which allowed large sheets of flat glass to be manufactured, had not been invented. The frames can be very slim as the glass is light and the sashes hang evenly balanced from ropes without the stress of a side-hung casement. The reveals were often splayed and internal shutters were common. In their original form these windows are generally seen as beautiful, and planning control may not allow replacement with anything that looks different. Unfortunately, they are cold and draughty and may let driving rain in. Trying to keep the look while upgrading the frames and glazing to modern standards can result in an ugly compromise.

During the twentieth century, hinged casement windows had a similarly well-thought-out rationale. Float glass allowed us to dispense with the very small panes and we have a small top-hung casement to try and provide background ventilation. A larger side hung casement is for purge ventilation and escape and a larger fixed pane provides the bulk of the daylight and a framed view out. This arrangement can be quite visually attractive if made from thin steel frames but starts to get quite clunky with modern timber or UPVC frame sections, which can easily take up over half the window area.

The use of trickle vents and fans in the twenty-first century, allowed the small casement to be omitted. A smaller side-hung outward opening casement permits purge

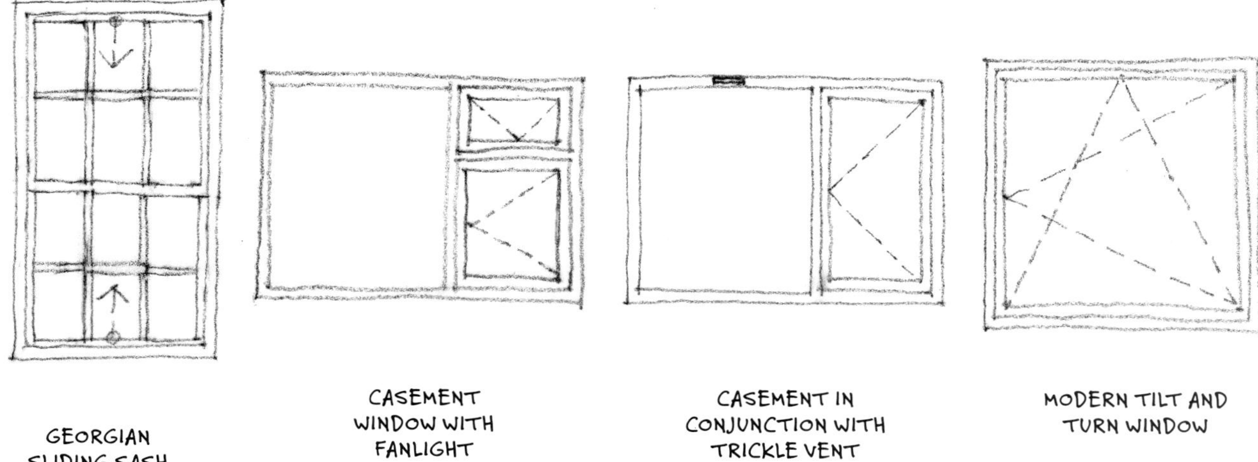

ventilation and escape, but the width of the opener is limited because of wind load and the risk of walking into it if at ground level. This is solved by providing a larger fixed pane for extra daylight.

Bringing us up to date, we now have inward-opening tilt and turn windows, more common in mainland Europe but widely adopted for low-energy and Passivhaus buildings. The window hardware allows a larger casement to be used, and tilting the window provides secure draught-free ventilation without the risk of the wind taking it off its hinges. The casement can still be opened inwards for escape or cleaning, or even for ventilation if it opens against an internal wall to avoid the risk of walking into it.

Inward-opening windows evolved in climates with hotter summers as they allow the use of external shutters to shade out the sun, while allowing secure night ventilation for cooling. Keeping wind-driven rain out was made easier with outward opening storm-proof casements. Of course, modern inward-opening windows are more wind- and weather-tight than traditional outward-opening versions.

For a triple-glazed tilt and turn window, the largest practical size is about 1.2 metres square (about 4 foot square). If a larger glass area is required, the layout in the third example can be adopted or a separate window can be installed. If the window is made only slightly larger, the extra mullion and frame thickness of the opener can actually reduce the useful glass area, while also increasing the cost. For that reason, there are increasing examples of buildings with simple tilt and turn single casements up to about this size. Usually, this is plenty for a normal-sized bedroom, bathroom or small office. For a larger non-domestic building, such as a school or health centre, such windows are used for individual staff offices or medical consulting rooms. Making the window narrower allows it to be taller without straining the hinges or the user. This is less optimal in terms of frame factor (the proportion of glass relative to the frame area) but it allows a lower sill for low-angle views or a higher head for improved daylight penetration.

Slate House, a cost-effective Passivhaus (see page 144), has single pane windows with no transoms or mullions. Fixed windows cost 30% less than opening ones, so the larger windows are fixed. Every room has at least one opening window and there is plenty of opportunity for cross-ventilation for night cooling in summer. Size and position are dictated by views and daylight. This gives clean, framed views and more glazing than the client thought they could afford. A canopy provides shelter and shading for the garden door and largest south window.

Windows and external doors

Softwood was widely used in the 1960s and 70s in the UK, but the frames were often badly designed and poor-quality wood was used, giving softwood windows a bad reputation. Modern timber windows are better designed, more robust and use higher quality timber. They are further protected by modern preservatives, paints and stains, and vulnerable sills and bottom beads are now typically made of aluminium or other durable materials.

The other principal material for high-performance windows and doors is PVC. This is an organochlorine compound, and as such is responsible for emissions to the environment during manufacture and disposal of extremely toxic and highly persistent organochlorine compounds. The use of PVC for building and many other applications has raised huge amounts of controversy. There have been a variety of reports, taking into account longevity, maintenance costs, the environmental impact of aluminium cladding and forestry practice and so on; none of which seem to be able to conclusively answer the question of the true environmental impacts involved. However, many environmental organisations have campaigned against PVC at times in the last decade or more. Jon has followed his preference for timber and has found that detailing timber windows with aluminium sills, set back in the openings with deep roof overhangs and good-quality stain recoated every 12 years, has shown no signs of deterioration.

There have been advances in window design and technology using other materials such as glass-reinforced plastic to create strong rigid insulated frames, which can be built into wall openings to hide the frame completely. This achieves continuity of insulation between the glass and the wall, reducing the need for maintenance and maximising daylight.

Glazing for low-energy buildings

A few years ago, the energy saving payback on triple-glazed windows was questionable in the UK climate. At the time of writing, some suppliers now provide triple glazing for the same price as double. Indeed, for some European manufacturers double glazing would be considered as special and would cost more. Triple glazing provides immediate benefits in improved comfort due to warmer surface temperatures in winter. In climates such as the UK, triple glazing is required in a Passivhaus building for this reason, not for the energy saving. While the glass will still be cool to the touch, we can sit close to it on the coldest of days without discomfort due to low radiant temperature or downdraughts.

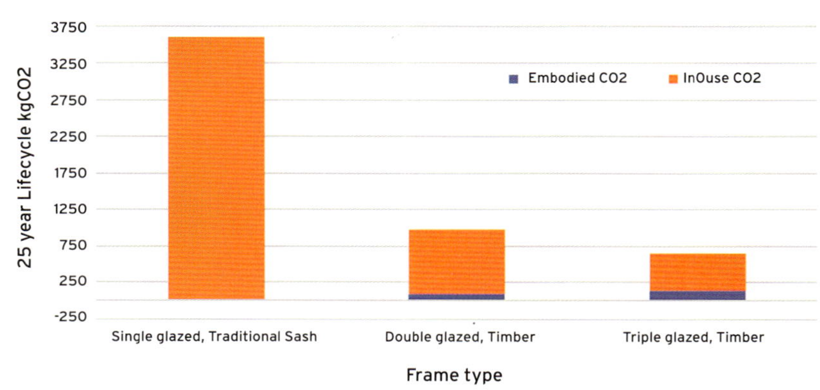

LEFT Triple glazing has slightly higher up-front emissions but design optimisation can reduce cost and upfront emissions. Graph by Mark Siddall, LEAP.

Some detractors argue that the extra pane of glass and argon gas require more carbon than is saved. (The argument being that carbon emissions avoided now are more important than the savings over the next 25 years or so.) If you are serious about reducing the upfront carbon of glazing – and we should be – then the best approach is to optimise the amount of glazing. We can avoid aluminium frames and prolong the life of the window and sealed units with good detailing, which also reduces heat loss and condensation risk. While all habitable rooms should have opening windows, not all need to open. A simple fixed glazing unit has less hardware and is significantly cheaper than an opening window, saving money and embodied impacts. As with building form factor, optimisation is the best way to reduce operational and, in particular, upfront carbon emissions.

Detailing around windows and doors

Detailing the head of a window right up under the ceiling will reflect daylight off the ceiling into the back of a room. The same applies for detailing a window hard up against a wall. These measures have the effect of avoiding a sharp contrast between the window and the wall surrounding the window, thus reducing glare and creating a graded transition from the brightness of the sky and the low light at the back of the room. It can be difficult sometimes to avoid a solid wall above a window to incorporate a beam over the opening, for instance. In this case, one can create a deep angled reveal to the opening, which will have a brightness midway between the bright window and the dark wall.

Other detailing issues around openings include keeping the weather out, particularly at the head and sill, while maintaining ventilation and drainage to any cavities in the wall or behind cladding and at the corner of the sill where it meets the window jamb. It is also necessary to maintain continuity of the airtightness barrier and insulation between the window and the wall.

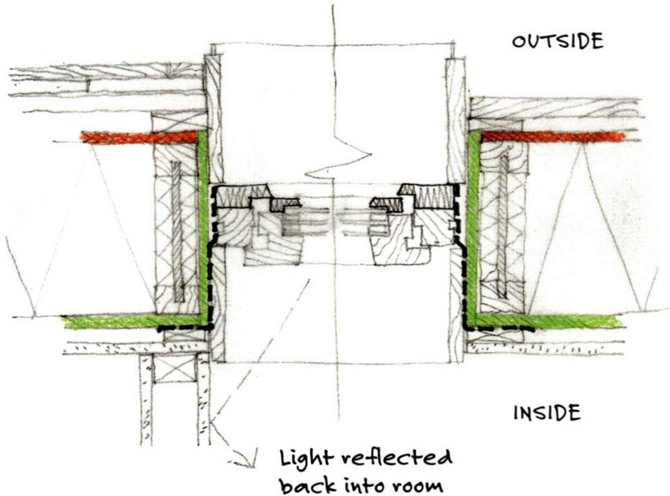

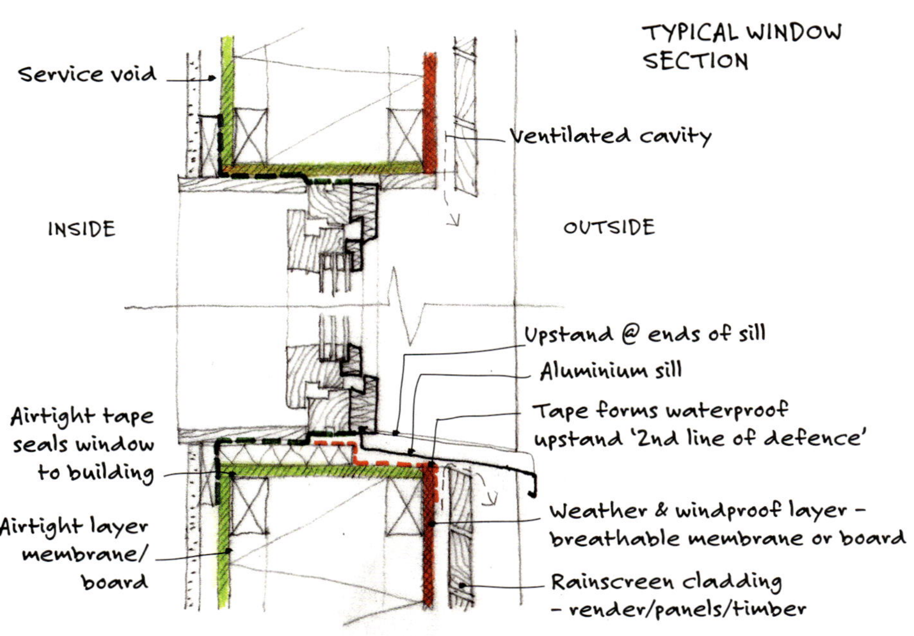

TYPICAL WINDOW SECTION
Service void
Ventilated cavity
INSIDE
OUTSIDE
Upstand @ ends of sill
Aluminium sill
Airtight tape seals window to building
Tape forms waterproof upstand '2nd line of defence'
Airtight layer membrane/ board
Weather & windproof layer – breathable membrane or board
Rainscreen cladding – render/panels/timber

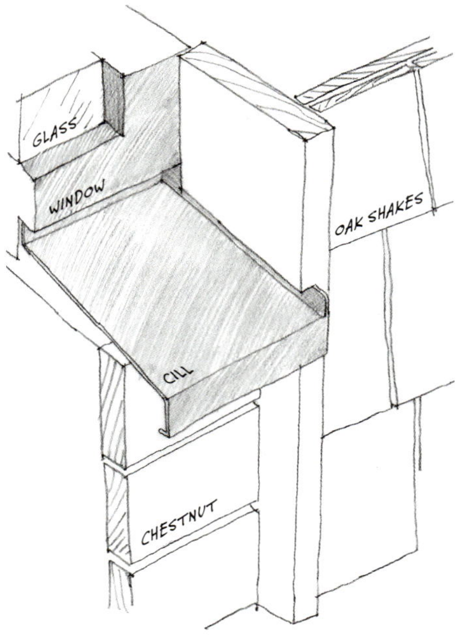

GLASS
WINDOW
OAK SHAKES
CILL
CHESTNUT

Services

LEFT Services as architecture at the Pompidou Centre in Paris. Piano & Rogers 1977. Exposure? Access? Not a good idea.

The mechanical services in a building typically comprise: heating, cooling, ventilation, hot and cold water and drainage. On the electrical side we have: power circuits, lighting, data and fire and security systems. The amount and complexity of each will vary according to building type and design.

As a rule of thumb, systems to provide these basic requirements tend to cost around one-third of the capital cost of a building and, after staff, are the largest part of the running costs of occupying a building, once the finance to buy it has been paid off. Glenn Hawkins' *Rules of Thumb*[1] specifies the following percentages of total construction costs for building services for different building types:

Building type	Building services as % of total construction cost
Warehouses	21
Flats/apartments	23
Offices	30-34
Primary schools	31
Clinics, group practice surgeries	35
Restaurants	42
Hospitals	45

Compared with a soundly constructed building, even the highest quality services will have a much shorter life. Kit wears out, becomes uneconomic to repair or sometimes even obsolete due to new technology.

The building services engineer, Alan Clarke, makes the case that for a good building the boundary between the fabric and services design gets blurred. Rather than coming in at a late stage to solve the problems created by the rest of the design team, early input can reduce or even eliminate the need for some building services and their associated cost and energy use. This is not how the industry tends to work. Where we see a move towards this integrated approach is when there is a clear and challenging performance target such as Passivhaus. Here, the heating demand and ventilation rates are known from day one and the building is designed to those clear constraints.

We will know with some confidence the plant size for heating, cooling and ventilation as soon as we know how big the building is and the number of people who will occupy it. We could be talking a house or large school (a hospital or factory will have more specific requirements).

Crucially, there is a pool of existing buildings that have already been built to the same standard. These buildings have years of post-occupancy data and can often be visited. Our example is for Passivhaus buildings but the principle applies to any evidence-based approach (see Peter Hazard's recent article for *CIBSE Journal*[2]). The lack of integrated design and closing the feedback loop is probably unique to the construction industry.

Second boilers for schools

The first Passivhaus schools in the UK used a small wall-hung boiler for heating, but as was standard for schools, a backup was provided so that kids didn't need to be sent home if it failed on a cold day. After the first winter, it became clear from the small temperature drop when the heating was off over the weekend that a backup boiler was not required. The next school had a single boiler; once the kids were in school they provided the heat for the building. Our first Passivhaus-certified archive repository was based on a small passive Danish archive with no air conditioning. As the client didn't quite believe it would work, tens of thousands of pounds were spent on plant that has never been used and on large ducts and service risers that waste valuable space.

Adding an extra gas boiler to a school is not a significant cost, but it does complicate the system design and controls. Now that we use heat pumps instead of cheap boilers, the avoidance of duplicate plant or backup heating for extreme weather is much more significant in terms of upfront cost and carbon. Unforeseen and unintended consequences are the bane of complex building services systems and controls. We can't foresee the unforeseeable but anything we don't install can't go wrong.

Building management systems

We have a love-hate relationship with building management systems (BMS). Once limited to non-domestic buildings, now domestic versions take the form of smart thermostats and home controls. These whole building controls systems are difficult to avoid in a building with complex building services yet, at least in our experience, they rarely work as intended.

The promise is that energy can be saved by integrating the control of window opening, ventilation, heating and cooling. The reality tends to be very different. For a start, if we get

the building fabric right then our buildings might cost less to heat than the annual support and maintenance cost for the BMS. That it will save any energy is a flawed assumption – even if the system could actually save energy, there isn't much to save.

Even the simplest software is difficult to debug. BMS systems tend to be bespoke because almost every building is different. The requirements vary by season, so commissioning typically takes over a year, but limited budgets usually mean that contractors want to do one big snagging session at the end of the defects period rather than fine-tuning with the seasons as issues emerge. Often fixing one issue reveals another, so this end of defects approach is unsatisfactory.

With complex building systems that are in effect always a prototype, problems are inevitable, people and technology are fallible and mistakes will happen. Complex controls are unavoidable these days, but where they are built into a heat pump or ventilation system by the manufacturer, we can expect a lot less trouble. This is because the same controls will have been developed and tested on multiple buildings over many years. Problems will still occur, but the manufacturer will fix them or go bust. By contrast, problems with bespoke whole building controls may never be resolved because it will never be clear whether the responsibility is with the user, architect, M&E engineers, energy consultants, the main contractor or of one of several specialist subcontractors or equipment suppliers.

The good news is that these problems can be fixed, but it needs a different approach.

Integrated design

We see several ways that the approach to building services design can undermine the quest for a good building. The first is an overly conservative approach of doing what has been done before. This may lead to a very competitive fee quote; it is easier and cheaper to copy and paste designs and specifications. In other sections we make the case for reusing design approaches that work and building on the past with incremental improvements. This is a good approach until there is a step change in technology or constraints. A tenfold reduction in a building's heating or cooling load gives opportunities for radical simplification of services, but that means new ways of doing things and consequently potential risks arising from innovation.

Over-design is another issue. This can be driven by fear of litigation – what if the building does not heat up quickly enough from cold? What if everyone in the block of flats wants to shower at the same time? Designing for worst-case scenarios has a capital cost and is likely to require additional plant-room space and perhaps a new substation or other upgrade to infrastructure. Most designers will follow guidance codified in standards that have served the industry well for many years. However, as we address heightened concerns about energy efficiency and summer comfort, some of these codes need to be challenged. The other reason for over-design is simply our tendency to overcomplicate things. On one project we were trying to design out the need for any cooling but the engineers really wanted to use a solar adsorption chiller as an innovative solution to gain BREEAM points.

This leads to the issue of justifying the fee. Again, few would openly admit to making things more complicated to justify the fee, but it is hard to avoid. Our second new-build archive store built on the lessons of the first, which allowed considerable simplification of what was already very simple by usual standards. When the engineer presented the design for the humidity control to the design team and client, the schematic showed three rectangles to represent the package chiller, a mixing valve and a cooling coil with three lines representing pipes connecting them together. The engineer looked embarrassed and said it felt hard to justify the fee before realising that the mixing valve was not required! As the design progressed, some complication crept in with the design of a very clever heat exchanger to save energy. This helped to 'justify the fee' but, unfortunately, the extra complication had some unintended consequences so the savings were less than expected. The next project embraced even greater simplicity with a single off-the-shelf ducted dehumidifier at well under one-twentieth of the cost.

The thought and experience needed to get to such a simple solution is less visible than a plant room crammed with kit and it can feel like we are short-changing the client when in fact we have saved a significant proportion of project cost and ongoing energy use. Ideas of engineers being paid a proportion of the saving in capital and running costs should encourage better design, but it is hard to imagine how such an approach could be made to work in reality.

We have used the example of archive building design throughout the book as a good example of a clear design problem with a simple solution, yet many of these buildings fail to perform because of their design. We see a similar story with the building services. Again, archive buildings may seem very niche but they make the point well. Obviously, all ancient artefacts have spent most of their life outside a modern air-conditioned store. Unfortunately, *when* and not if the air conditioning fails, the environment can deteriorate very quickly, causing damage to objects. General storage for the bulk of museum and archive objects requires an average temperature of around 18°C and relative humidity in the 40–60% range. What is well documented is that switching the plant off completely, usually results in improved conditions with little or no energy use. The air conditioning plant can take up 20–30% of the building floor area, costs a significant proportion of the building capital cost and is usually responsible for most of the building's energy use, and yet the building may work better without it. This is hard to believe but easy to prove, literally at the flick of a switch.

For other building types, the reduction and simplification in building services is generally less dramatic than for archives but the same principles of getting the building to do as much of the work as possible to achieve the desired conditions applies.

Hot water

Anything we might write about the technology of hot water generation and storage will be out of date very soon, but early decisions on building layout will have a significant impact on hot water consumption and the cost of installation of hot water systems.

For low-energy housing, energy use for hot water can be higher than for space heating. For larger buildings, storage and distribution losses can more than double the energy use of hot water systems and also contribute to summer overheating. We can specify regulated tap outlets and shower heads that deliver sufficient but not excessive flow rates but how much hot water people choose to use is mostly beyond the control of the building designers. What we can influence is the hot water system losses.

As an example, small-bore pipes have been used for hot water feeds to taps on projects ranging from homes to schools. This saves water and energy, and as the pipe comes on a coil it can be quickly run through service voids with no inaccessible joints that could leak. However, concerns that the user will complain that the flow rate is insufficient can lead to installers increasing the pipe size. This wastes hot water and users have to wait longer for water to run hot. This may be countered by adding a pump and recirculation eliminating the wait but greatly increasing losses. For larger non-domestic buildings such as schools and offices, the water usage per person is much lower than for housing and so distribution losses dominate. Larger buildings require hot water to continually circulate to reduce dead legs. Because of this, it is usually cheaper and more efficient to use local, direct electric hot water supply rather than centralised hot water storage heated by a gas boiler or heat pump. The storage and distribution losses negate most of the theoretical energy savings of a heat pump and the losses can make a significant contribution to summer overheating.

Two of the first Passivhaus schools in the UK used efficient gas boilers and carefully insulated pipes with a timed circulation pump that only ran during occupied hours. Monitoring showed that losses accounted for around 70% of gas use for hot water. The next projects used local electric water heaters. Instantaneous heaters would have eliminated the need for storage but were deemed unsuitable because of the potential peak current requirement during break times. Instead, small stored hot water units were used but with each one feeding several outlets via 8mm copper pipes. This required some design effort but installation was much cheaper and easier than a traditional system. The installer was sceptical about the small-bore pipes so a mock-up was made by the contractor at no extra cost, in the expectation that it would not work. However, it proved to be a workable and economical solution.

The hot water system rectangle

Gary Klein in Sacramento, USA, has devoted many years to teaching designers how to optimise hot water distribution. He is an advocate of small-bore radial plumbing and optimised plumbing layouts and knows how difficult it is to get designers excited about such apparently mundane topics that can make our buildings measurably better. One idea he has developed is the *hot water system rectangle*. This can be embraced by architects at the early design stage before the building services engineers are involved. The designer draws a rectangle on the floor plan that includes the hot water source and all the hot water outlets. The area of this rectangle is then divided by the gross internal floor area of the building's conditioned space to give a percentage.

The challenge is to get this figure as low as possible. Often, large reductions can be achieved without changing floor plans but by simply flipping where plumbing fixtures are located. More radical rethinking can get this figure under 1%. Although the design tool is a

ABOVE The contractor built this rig to prove that the specified small bore copper pipe to taps would not allow enough flow rate!

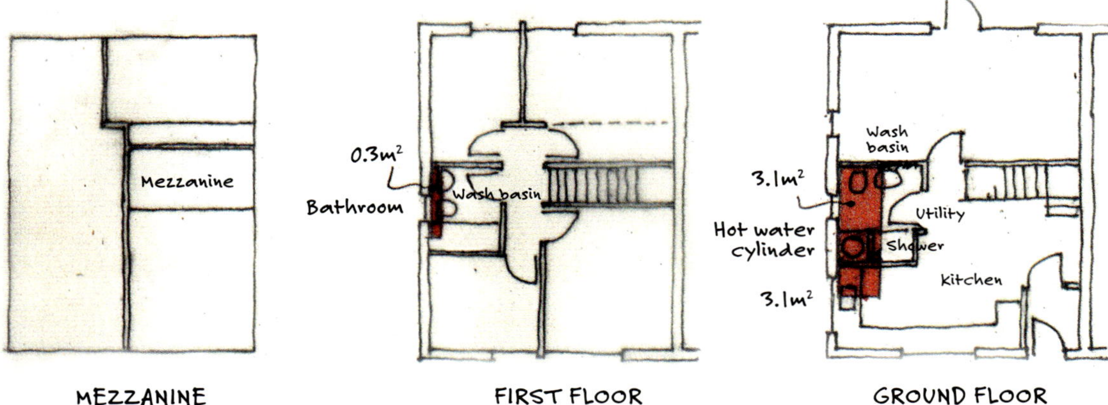

ABOVE An efficient layout for a house demonstrating the hot water rectangle developed by Gary Klein in Sacramento CA.

new idea, planning around a service core was the normal approach when plumbing and drainage was seen as an expensive luxury.

Optimised plumbing layouts will dramatically reduce hot water losses and reduce wait times for taps and showers to reach temperature but they will also reduce the cost and upfront emissions associated with drainage. Simpler layouts greatly reduce the risk of unforeseen complications and expenses as the designs progress. Examples include drain runs clashing with structure or the need to achieve suitable falls and rodding access within floor depths. Further simplifications include simplified ventilation layouts and the potential for easier access to plumbing and drain stacks for maintenance. A simple stacked plumbing core also greatly reduces the risk of damage should a leak occur.

This is a good example of a creative design approach using clear constraints and metrics for something that is not usually quantified. The result should make a better building with improved comfort and function that is cheaper to build and run. However, the design effort will be invisible to all but the most trained eye.

Services bling

It is easier to get excited about more visible technical fixes, but just as green-looking architecture such as living walls or passive solar design usually fails to reduce environmental impact, the same is true of building services eco-bling.

An energy transition is underway and is likely to be undertaken within a very short timescale. This implies key impacts on the design of buildings; heat pumps provide space and hot water heating. Early experiences of very high electricity use led to the belief that heat pumps are only suitable for very efficient buildings. While reducing heat demand is crucial to lower the pressure on a renewable grid in winter, there is now a growing body of evidence that properly designed heat pumps can even work well in existing uninsulated buildings as long as they are run continuously and at a low flow temperature (see Chapter 2.6). Some heat pumps now employ propane as the refrigerant, which has an extremely low global warming potential of less than 3, compared to around 750 for modern alternatives (although the legal limit is set to reduce to 150). Meanwhile, the cost of grid-linked photovoltaic panels has dropped significantly, but the question remains as to whether individual systems on buildings are the best use of resources.

Finally, there are some basic principles regarding services:

- Get the building to do most of the work of keeping warm in winter and cool in summer.
- Avoid oversizing ventilation, heating and cooling systems.
- Consider plumbing design in-depth when planning layouts to minimise costly drainage runs, vent stack and hot water pipe runs by grouping rooms that need these. This gets even more important as buildings get bigger.
- Keep things as simple as possible but no simpler. Mechanical ventilation provides controlled background vent whether windows are open or shut and is more effective than so-called 'natural' ventilation.
- Design the ventilation at an early stage, especially for multifamily and non-domestic buildings. Minimise cold air duct lengths (to and from outside).
- Consider window ventilation complementary to the mechanical vent strategy for night cooling, purge ventilation and occasional peak occupancy.
- Plan space for plant and routes for wires, pipes and ducts in walls and floors; connect without prejudicing structural members.
- Do not skimp on specialist design input. A good services engineer will save their fee in reduced capital cost by designing simpler systems.
- Do not cut corners on quality. Ventilation systems need to be efficient, silent and draught free if they are not to be switched off. Plumbing fitting and sanitaryware need to be durable and repairable if it they are not to leak.
- Allow good access to services for repairs, maintenance and adaptability. This is not easy with a preference for services to be concealed in walls, floors and ceiling voids.
- Avoid complex digital controls, if possible, as many people find them difficult to use. Even large buildings can avoid BMS.

TOP A cheap and efficient mini-split unit has replaced the woodstove in Old Holloway Passivhaus with a single indoor unit in the kitchen. This is a popular option in climates that also require cooling. Hot water is provided by an integrated air to water heat pump.

ABOVE And the external unit.

Edwards Court extra care council homes in Exeter used precast concrete panels and brick cladding with full fill mineral wool cavity insulation and is the first Passivhaus certified extra care facility within the UK. Architype 2022.

4.0 The business of building

4.1 Land Focusing on the issues of land ownership and value, which have huge influence on what is built

4.2 Design education and post-occupancy evaluation How to acquire the skills, knowledge and values designers need; learning what works and what doesn't

4.3 Cost and value Why do costs vary? Who benefits from value? How to reduce uncertainty and control costs

4.4 Value engineering in design Understanding how buildings perform and learning from experience

4.5 Appointing a design team and contractors A constructive approach to finding the right team and managing costs, quality and time

4.6 Risk management and regulation Risks that cannot be avoided and how to mitigate them, and the uses and limitations of regulation

Land

The ownership of land in the UK is uniquely concentrated compared to elsewhere in Europe (most of the work on investigating land ownership has been carried out in England and so data for Scotland, Wales and Northern Ireland is limited). This leads to a number of problems, including inadequate access to comfortable and affordable homes, ecosystems with declining biodiversity and poor responses to climate change. Originally, land was held in common. Anyone could come and go where they wished to obtain food and satisfy other needs. There was no private ownership. It was only when someone put up a fence around a patch of land and made it theirs alone that the problems started.

It is necessary to go back to 1066 to understand the background to this and consider how one might improve the situation. William the Conqueror took over the whole country for the Crown and proceeded to gift half the country to 200 or so of his barons in return for their loyalty. Remarkably, nearly 1,000 years later, it is estimated by Guy Shrubsole in his book *Who Owns England?*[1] that descendants of around 4,000 peers, baronets and gentry still own around 30% of England.

Much of this is upland grouse moors but it also includes urban estates, including the four 'great' estates that own nearly all of the West End of London, estimated to be worth around £20 billion. Nevertheless, the Crown still controls half a million acres and together with that other pillar of the establishment, the Church, they account for approximately another 2% of the land in England. This is about the same as the total owned by the National Trust, RSPB and other conservation charities all together. The public sector – including the Ministry of Defence, Forestry Commission, local authorities, roads, railways, schools, colleges, hospitals and other government departments – own 8.5%. The private sector, corporations and wealthy individuals from home and abroad own a whacking 35%. Far from being a 'home-owning democracy', the population as a whole lives on just 5%. The aristocracy plus around 20,000 wealthy individuals at home and abroad (that is a tiny proportion of the population who make up around 0.04%) own a total of 50%!

As long as what happens on the land, how it is used and for whose benefit is governed by a few, there will never be an England that reflects the values and needs of anything but a minority of its citizens.

Problems of land ownership

This leaves 17.5% of England, the ownership of which is uncertain as it is not in the public domain. This too is remarkable as just about every other country has cadastral maps of every parcel of land and who owns it. Transparency is vital to address some of the issues arising from the phenomenally unequal ownership of land in the UK. The ownership of agricultural land attracts substantial public subsidy, according to Guy Shrubsole: £8.4m in 2015 to the 24 non-royal dukes alone. Meanwhile, subsidies directed towards planting trees would reduce flooding, and rewilding would improve biodiversity.

Vesting ownership of land in overseas tax havens and avoidance of inheritance tax has exacerbated the inequalities. It has also made land ownership in the UK attractive for money laundering by crooks and oligarchs. Landowners have benefited from unearned gains in value from stealing common land over the centuries to more recent windfall gains from development. For example, the building of the extension to the Jubilee underground line in London cost the public purse £3.4 billion, but the increase in the value of land within 1,000 yards of the 11 new stations was estimated to be £13 billion and this went straight into the pockets of the landowners. Developers have secret options on potentially developable tracts of green belt surrounding many towns and cities and this drives up the price of any parcel that is capable of having planning permission granted.

The housing crisis

There are a number of aspects to the crisis in housing; existing homes that are in very poor condition, in the wrong places and unaffordable for much of the population. In addition, not enough homes are being built to house the rising population. While there are around 800,000 homes standing empty – second homes, holiday lets or just empty – it is estimated that there is a net requirement to build 300,000 new homes per annum to relieve the current critical problems. For this reason, the housing crisis is a land crisis.

The price of agricultural land in the UK has increased by 300% in the last 25 years. Meanwhile, the UK has never implemented land reform and attempts to tax windfall gains failed after both the First and Second World Wars. The ability for local authorities to purchase land at its existing use value underpinned the rebuilding of Britain in the 1950s and 60s until this was repealed in 1961. The situation has been made much worse because 5 million acres of public land worth £400 billion has been sold over the last 40 years as part of 'reducing the state', followed by the 'austerity to restore public finances' agenda. Unfortunately, this realised only a fraction of its worth for reinvestment in public infrastructure, Meanwhile, it has eliminated the possibility of keeping it in trust for the public good, underpinning the provision of affordable homes and creating income for other public investment in the future. Furthermore, Rowan Moore in *Property: The Myth That Built the World*[2] argues that the shift towards our homes being treated as financial assets is why we are struggling to kickstart climate programmes that might improve their quality, liveability and safety.

Planning control in the UK

There is an assumption in the UK that ownership allows a more or less free hand with how land is used and traded. We have inherited a situation where landowners have very few restrictions on what they can and can't do on their land, without any of the responsibilities that came with ownership in times gone by. This is in spite of land use and development being subject to planning regulation since the Town and Country Planning Act 1947. This aimed to bring development decisions under democratic control to achieve social objectives of good transport, shops, schools and other shared facilities. Local authorities are enabled to charge taxes (CIL and Section 106 payments) to mitigate the impact of a development on the local community and infrastructure. However, these taxes, which may be levied to fund schools or traffic management, for example, do not come close to reflecting the increase in value imparted by the grant of planning permission and do not generate a significant contribution towards the costs associated with providing good public infrastructure to support development.

Currently, planning control tends to concern itself with a great deal of detail, which imposes a significant burden on applicants. This is an irritant to commercial developers

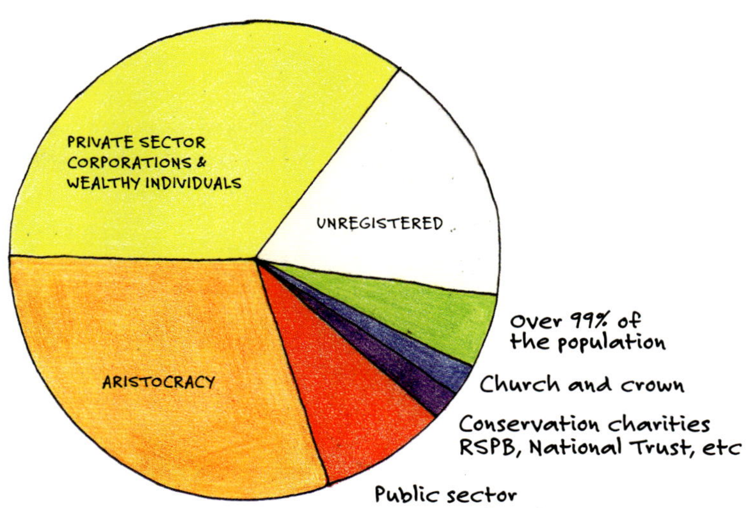

ABOVE Land ownership in England. 99% of the population occupy the green bit!!

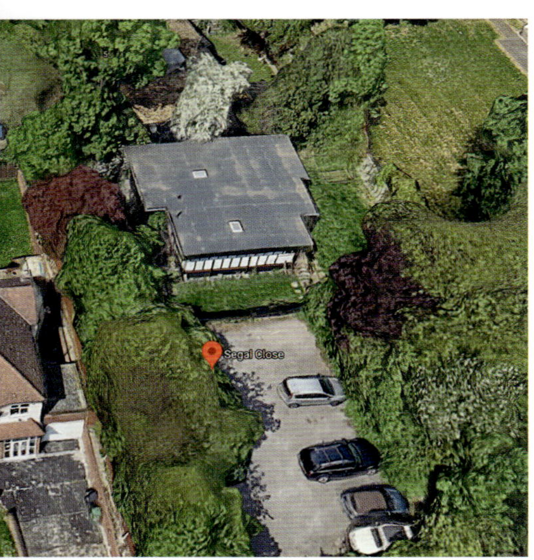

ABOVE A steeply sloping site with poor access in South London.

who have to factor time and money into complying with requirements, but it renders the whole process extremely difficult for individuals and community-based organisations to access. A recent application for a community-led project in London required the submission of 21 reports commissioned from experts, and responses to 39 detailed conditions on the permission. This required employing planning and other consultants; the whole exercise cost a six-figure sum to comply with. Currently, the average cost and time for a planning application referred to the Greater London Authority that is over 150 dwellings or of strategic importance is £125,000 and 12 months. Commercial developers are also able to apply significant pressure on officials and elected representatives using their financial clout to get what they want and minimise social benefits.

Finding a site for self-build and community projects

In the face of an unreformed market in land and a discriminatory planning system, what options are open to obtain access to land? The main consideration is that with a constricted market you should put access to a site first and adopt relatively open selection criteria rather than setting out a brief for an ideal plot in terms of location, size, ground conditions, access, services and so on. Remember that many urban plots are undeveloped for a good reason.

You can purchase land without planning permission at a significant discount and reduced risk of ending up with a site for which you cannot obtain a planning permission by making a conditional offer. Contracts of sale are exchanged but not completed unless you obtain the permission for the building you wish to build.

The situation is very different in rural and urban areas. It is still possible to find plots for building a house in rural areas or to create such an opportunity by getting planning permission to subdivide a garden, for instance. Rural exception sites are small patches of agricultural land outside a village boundary that would not normally get planning permission for housing. The planning authority agree to a landowner providing land at below full market value on the basis that the land is used to build affordable homes for local people. Alternatively, opportunities exist to develop difficult-to-develop sites, which are too steep or constrained in other ways, to make them attractive for commercial developers, although these days any scrap of land in successful urban areas is seen as an opportunity by developers or individuals wanting to build a one-off single house.

It is possible to identify land that is not advertised on the market that an owner may be willing to sell by getting on your bike and travelling the streets with a 1:1250 scale map and/or a Google satellite view. Follow up anything that looks promising by checking who owns it by investing £30 in a search at the Land Registry. This will get you the details of ownership. There are a surprising number of parcels of land with no obvious use.

The Church has significant land holdings. Much of it is administered by the Church Commissioners whose remit is to generate income for remuneration and upkeep. However, there is a growing sentiment that the Church's resources should be better directed towards social aims.

Another possibility is to enter into a partnership arrangement either with a housing association with similar social aims or a developer for whom you can fulfil their responsibility to provide affordable homes under a planning agreement. Remember that in either case your ability to control the major decision will probably be reduced.

Land from a local authority

Another approach is to see if the local authority is willing to make land available. Each local authority is required to keep a 'Right to Build' register of people and groups in their area who want to have a role in an owner-commissioned home, whether that be for custom building, self-building or through community-led housing, such as cohousing or CLTs. They also need to give suitable development permission for serviced plots to meet this demand. However, at the time of writing, most authorities have not yet met this requirement and some have gone further by introducing requirements for a local connection or excessive fees to reduce the authority's responsibilities.

Local authorities are required to obtain 'best value' when disposing of assets and this is generally understood by cash-strapped councils to mean highest bid. The rules do allow councils to take 'social value' into account through the Public Services (Social Value) Act 2012, although little use of this provision has been made in the creation of affordable homes. There is a body of academic work devising ways of measuring social value which could be creating affordable homes or community facilities and this permits a site to be given to a socially beneficial project at little or no cost.

You will need to persuade relevant council officers in the housing, planning and valuation departments as well as gain the support of elected members. You will probably not have a track record but you must appear credible. Make use of the local media and recruit a significant membership. Do your homework: who are the relevant personalities, mayor and ward councillors in the area of a potential site, and what are the relevant council policies regarding housing strategy, local plans and disposals? You may need the support of the local MP. You need a clear ask (how much control do you want over the development) and a clear offer; for a group this could be genuinely affordable, high-quality, low-energy homes forming a new sustainable neighbourhood. If you are a community-led organisation you may be able to bring in finance for housing into an area that would not be available otherwise. You can offer homes to existing social housing tenants in the knowledge that any home made available would be offered to a household at the top of the local housing waiting list.

In urban areas, there may be possibilities to develop within an area of social housing that is being regenerated. This may involve the refurbishment of some buildings coupled with the demolition of others or new buildings constructed around the existing buildings to increase the density. A community-led development could be beneficial in offering opportunities for new affordable homes and homes for existing residents as well as working with existing residents to create new landscapes and other facilities. These measures require dedication and resolve in the face of disappointment.

Stop sharing
L5
P12 L2
P11
S3
Ecococon panel drawing

Design education and POE

The necessary skills include procedures to solve design problems as well as drawing, making and digital literacy. Architects need an understanding of building construction and materials and structural and environmental design as well as the business of building: regulation, risk and liability. They need awareness of history, social values and the existential crisis facing mankind. Design education will often involve some unlearning of preconceptions and prejudices.

Bashar Al Shawa writing on a blog regarding architectural education said:[1]

> *Let's start by agreeing that, at the end of the day, architecture is supposed to make us good buildings and that architectural education should teach students how to make those good buildings? Sadly, as I experienced it, and I don't think that architecture schools are much different, it almost succeeded in teaching me how to make bad buildings.*

Both Nick and Jon were taught a systematic problem-solving approach: understand the problem/brief and the context/site then make a series of propositions/tests/analysis/improvements and repeat. We also both have experience working on the factory floor or building sites alongside those responsible for the work and we have built our own homes. Although this was some time ago, the experience has a big influence on our work today.

Design has gone through recent preoccupations with futuristic visions of a technological wonderland or a mystical dome world: buildings with pipes and ducts as architecture, design as activism, postmodern anything goes and buildings as parametric sculpture. Student have to find their way through this mess and form their own conclusions with the help of their teachers.

We have seen new young students overwhelmed by the new world of abstract thinking they find themselves in. They often come up with an outlandish 'concept' that they are asked to 'develop', which they do without reference to the real world, ending up shoehorning the requirements of the brief into the form developed from their initial concept. At this point, some students become confused and depressed.

What designers need

Designers need to be able to think space, to visualise what it is like to be in a place, how spaces are animated by people and how they are affected by sound, light, colour and texture. This awareness needs to be developed at different scales; how a handle feels to touch, what it is like to be in a room, or a house to live in, how the street is outside and how a city is to inhabit. They need to be able to envision spaces at different scales and understand how to construct them.

They will need to communicate effectively, through words and drawings, first with clients to allow them to visualise what places might be like, second with regulators and planners to convince them that proposals are safe and compliant with building standards and planning rules, and finally with contractors to inform them so they are able to build the design.

They will also need to understand the principles of building physics as it defines the environmental and structural design of buildings and the materials they are made from.

Most importantly, designers should understand the needs of their clients and be able to create optimal solutions to meet their wishes, and not impose the needs of their individual egos. We have seen new young students overwhelmed by the new world of abstract thinking they find themselves in. They often come up with an outlandish 'concept' that they are asked to 'develop', which they do without reference to the real world, ending up shoehorning the requirements of the brief into the form developed from their initial concept.

Some thoughts on courses

Many courses are now based on some form of student-led learning; part-time practitioners lead 'units' with a variety of approaches. Students choose a unit and set their own projects. This models the realities of practice to some extent and requires students to be generally aware and also self-motivated, but support for less-able students is required.

However, it seems to us that the teaching of cost, structure, environment, construction and so on are not treated with sufficient rigour, or are ignored altogether and not applied in a consistent way to students' projects.

Building is a team effort involving designers and constructors as well as engineers, cost consultants and others. Students need to have sufficient knowledge of these fields to be able to understand the work of other members of the design team, to make suggestions to resolve problems and make improvements, but also to admit ignorance and ask for advice where needed. Sorting out issues on site with contractors and others is a critical part of creating good buildings.

We have both benefited from on-site experience, which develops an understanding of the nature of construction: materials and how they go together, their structural capacity, the importance of tolerances and other vital aspects of creating good buildings. It is important to develop mutual respect between designers and makers. However, it appears that opportunities to experience real-life site operations are much reduced; courses occasionally run live projects, often collaborating with local community groups as clients.

Such projects are generally very successful but sometimes difficult to organise within the framework of the academic timetable.

Some courses aim to bring actual design projects into the teaching process, with students developing projects in collaboration with real clients. Another approach has been to embed a working design practice within an architecture department, where students get an opportunity to work alongside their tutors on actual design projects. This provides experience in real world design and practice. Also, research undertaken in architecture departments has grown and forms a potential source of expertise to inform undergraduates.

Other courses offer experience in overseas working with local professionals in workshop sessions developing projects. This provides an opportunity to understand the social context and building culture of other places, thus opening up new ideas.

Some departments have well-equipped workshop facilities available to develop craft skills and modelmaking. The role of sketching and drawing in design is discussed in Chapter 1.6 and some courses provide opportunities for life drawing and outdoor sketching to develop these skills.

While energy-efficient design and sustainable construction are now established as necessary attributes of buildings, much as you would not design without taking structural safety into consideration, it seems to us that there are a number of areas such as building construction that deserve to have more priority than at present.

Refurbishment and the repurposing of existing buildings is more difficult than building from scratch and yet more important in achieving a less energy- and resource-intensive future. So too is designing for a long useful life and with consideration of repair and maintenance of buildings. Also related is designing for adaptability so that new buildings can be altered to suit changing needs.

Too often one sees designers presenting a new building as an object suspended in complete isolation from its surroundings. Context is everything; most new buildings are relatively insignificant in the context of the city and the constraints imposed by the surroundings as well as the social and financial context will direct the problem-solving process and ultimately determine a good solution.

The economic context of construction is crucial and yet does not seem to be considered in any detail in most architectural courses. The financial implications of the balance of capital with revenue expenditure and the qualifications imposed when raising finance can affect the design in fundamental ways. Some architectural courses are run in collaboration with estates management, planning and surveying courses to plug these gaps.

Finally, planning, building and fire regulation are significant constraints, and information on what they are intended to achieve and how they do it is important in order to be able to devise economic and practical solutions and to communicate their effectiveness to regulators.

We believe that learning by doing is the most effective approach: more showing and less telling. It is important to be empathetic to the students' point of view and to anticipate

ABOVE The architect Louis Kahn famously suggested asking a brick what it wants. We suggest that there may be much more to be gained from chatting with a bricklayer.

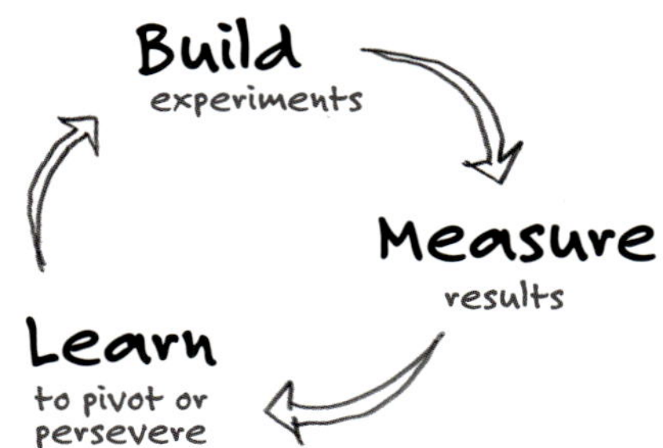

ABOVE Few designers or builders go looking for problems after handover but that is the best way to improve.

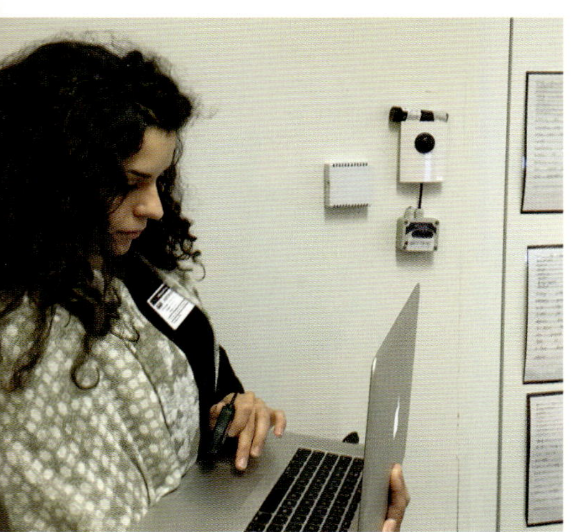

ABOVE Chryssa Thoua from Architype carrying out post-occupancy monitoring on a range of schools to evaluate air quality and comfort.

the potential difficulties they may have. Problems need to be set in their context and backed up by research and by doing the numbers to create a firm basis for moving forward.

You may conclude from the outline above of some measures to improve architectural education that courses would need to be extended to accommodate additional topics. We do not believe this to be so – indeed it is our view that full-time teaching could be reduced with more emphasis on learning on the job.

A recent development has been the first foundation of an independent school of architecture since 1847: The London School of Architecture, which offers a two-year course to RIBA Part 2 qualification to students who combine part-time work with studio learning. The aim is a practical education taught by a mix of practitioners and academics.

Also a perceived lack of in-depth technical teaching in structural design and environmental engineering in particular has led to a range of graduate courses with titles such as 'Technical Architecture' and 'Architecture with Engineering', which aim to fill this gap in understanding among designers. Nick's experience teaching low-energy building physics and detailing to a wide range of professionals suggests that architectural technologists usually have a far better grasp of how to put a building together than many architects.

Meanwhile, the Architectural Registration Board, which controls entry into the profession and regulates architectural education, has published wide-ranging proposals. These include competencies that will be expected of architects (49 of them – interestingly, competence in preparing digital information is one, but hand drawing is a skill not required by the regulator).

It remains to be seen how this will change architectural education, but it aims to open up new routes into the profession by not requiring an undergraduate degree in architecture, encouraging mature students and people from different backgrounds to join the profession, which is a good thing. It also aspires to 'make the approach to quality assurance more proportionate' which I think means that they agree that current assessments are too complex and prescriptive, making staff and students over-preoccupied with marking schedules, grades, *etc*. Our experience suggests that this limits imagination and absorbs a significant amount of staff resources.

Post-occupancy evaluation

Post-occupancy evaluation (POE) is the process of obtaining feedback on a building's performance. This should lead to immediate fixes for problems but also allows feedback and education to improve future buildings. This sounds like a good thing, so why does it not happen as a matter of course?

One problem raised is how to get the client to pay for this extra service. Architects and consultants argue that it will provide benefits that will pay for the investment. However, clients reasonably ask why they should pay extra to get the building to actually work?

When we buy or commission anything other than a building, we expect it to work and we have consumer rights that will ensure a defective product will be fixed. If the product fails to deliver what was promised, we can get our money back. This is why sellers of health and beauty products have to be so careful not to claim any benefits that cannot be verified.

Conversely, we find that most stuff we buy is pretty reliable these days. If it isn't, the company making it will lose money fixing the defects – most likely more than was earned selling the item. This is not a sustainable situation, so the product must evolve or be discontinued. The short-term exception is products that are so cheap relative to the hassle of getting a replacement that almost no one bothers to claim them.

The situation with buildings is different to most consumer products. As with health and beauty products, claims may be vague with few quantifiable metrics. The nature of buildings also means that there is plenty of room to blame performance shortcomings on the occupants. If a building overheats in summer, the designer might point out that the building was modelled and shown not to overheat as long as the occupants open the windows once indoor temperatures reach 22°C. It is only when outside temperatures are lower than this that the occupants and not the building are at fault. Also, not all occupants of the same buildings are complaining about the heat, so it can't be the building. The fact that the owners are sensitive to road noise and that flies are a particular problem locally is hardly something the building designer could have been expected to anticipate. If heating bills are high then it is again probably the occupants leaving windows open or choosing not to wrap up in winter.

In the UK at least, another key issue is the very low expectations we have for building performance, whether in terms of energy bills, comfort or even just keeping the rain out. Even if we do care about performance and quality and if money is no object, we will probably put all that aside for the right location.

On top of all this, most buildings are effectively one-offs, so we don't get to read 20 product reviews before parting with our money.

All this assumes we have the money and freedom to choose where we live.

If the designers and builders really care about the performance of their building, they will be doing a POE and seeing it as cheap research and development, rather than something to be charged to the client.

The good thing about this approach is that it is in the interest of the design team to make the building work with few problems. Fixing defects is very expensive in terms of time, money and the reputation of the designers and builders. We will also think twice about trying too many new ideas on each project. Rather than being driven by the excitement of the new, we will seek a deeper pleasure in fine-tuning layouts and details that we have been refining over many projects (see Chapter 1.4).

In a world with instant global communication, we have the potential to learn quickly from each other's mistakes. To be useful, this requires a no-blame culture, with everyone working towards a common goal. This might happen through the goodwill of people sharing lessons at conferences or on technical forums but also through the feedback of happy or dissatisfied customers on social media.

It is not easy to admit mistakes or to ask occupants for feedback, which could be negative. Indeed, our insurers warn that we should not admit any liability for defects without first contacting them.

ABOVE Data loggers for an in-depth monitoring project in schools. Energy bills and formal or informal questionnaires are also useful where there is no budget for formal POE.

Cost and value

The cost of construction is very variable (and generally more than you thought). It is affected by the shape and slope of the land and the nature of the ground, the complexity of the layout and standard of finishes, the specification of services and many other specific circumstances.

The cost can be difficult to anticipate – whether it's HS2 or your back extension. Even a simple project is affected by the price and availability of both materials and labour and the demands of the planning system. And that is leaving aside variations in overall costs arising from purchasing land, appointing a design team, other professionals and fees for permissions, *etc*. A building is not like a consumer item with a generally anticipated price and level of performance. It is subject to the context, and the price of a new building is also affected by the level of the commercial and residential property markets generally.

Rather than using this uncertainty as an excuse for projects always going over budget, we can look for ways to reduce this uncertainty. The pursuit of simplicity and building on what we know works are good ways to reduce risk. We look at other ways of reducing uncertainty in Chapter 4.6.

The following can all help to keep a project on budget:

- Paying the contractor early in the design process to bring their knowledge of the supply and labour markets to bear when estimating costs.
- Paying the contractor a fixed sum to 'design and build' the project, which means they include the risk of costs rising during the course of construction. You pay a price for the contractor taking the risk of costs rising and taking longer to build and so this will probably not be the cheapest price, but it does offer a higher (but by no means absolute) degree of cost certainty.
- Carrying out some of the work in a factory away from the weather can improve certainty but can limit choice, and is unlikely to be as cheap as site-based construction as factory overheads are generally greater than the potential savings. To get the full benefit you need a relatively standard product which enables you to establish a fixed supply chain and an established workforce trained for the particular job in hand.
- Developing a design for manufacture with every detail and component specified.

Why building costs vary

Walter Segal scheduled every component needed, sought competitive quotations for all materials required and worked with a trusted team of carpenters, which can work for a modest and relatively straightforward project. His schedule was an actual shopping list of items required to build the project rather than a general 'measure' of the various elements of the building that the contractor's surveyor then has to turn into a shopping list. Segal's list is used to obtain accurate competitive prices for all the items in the building at an early stage of the process.

Building costs are generally estimated from the costs of previous projects. The published figures are often the price quoted at the outset at tender stage. However, actual costs on completion are often higher. The figures are correlated to region, time and type of project which helps, but this aggregation reduces everything to an assumption of 'average complexity' and this masks the high level of variation that is present.

The question of ways to reduce building costs is addressed in the next chapter and focuses on increasing performance at no extra cost or reducing cost for no reduction in performance. Improving buildability or 'simpler than average complexity' of construction is a key way of reducing time and cost without a loss of functionality. This is something a builder will appreciate more than a designer or quantity surveyor (QS).

On a recent project, the structure did not quite align so a glulam beam was added to transfer the load. Not one but two beams were required. When the carpenters went to fit the first beam it was about 200mm too long so they cut it to length. Unfortunately, the second beam was 200mm too short! What should have been a minor increase in complexity and cost was compounded by a simple site error, leading to delays and further costs.

Ensure that the cost consultant is involved in design development from the outset and that they prepare estimates as the design progresses. This sounds like good common sense but it is surprising how a QS can be reluctant to engage fully until the design is relatively fixed, by which time it is often too late to make the fundamental changes that may be necessary.

A fundamental issue is examining the priorities in the brief; putting money into more complex forms or expensive finishes rather than useful floor space, as an example.

Another aspect of building costs is builders' profits. According to Colin Axon and Simon Roberts in their 2022 paper,[1] high prices in the housing market have enabled large house builders to extract increased profits; £70k per dwelling at 2019 prices. Certainly, labour and material prices have gone up but the size and performance of dwellings has remained largely unchanged. Land prices, on the other hand, which are often cited as the reason for high house prices, have remained broadly stable.

It is necessary to understand the relationship between cost and value at the outset. Who benefits from value; is it the owners or the users? Who sets the brief to get a balance of social, economic and environmental benefits at what level of cost and risk? This is not a quick and easy process but rather requires extensive dialogue among the stakeholders.

ABOVE Three phases of social housing by Jon Broome Architects for Greenoak Housing Association in Woking 2004, Normandy nr Guildford 2005 and Storrington nr Horsham 2009. For each phase cost were reduced but performance improved.

Controlling costs

The vital aim must be to set clear performance targets and make sure that they are carried through the development process. This is to ensure that cost and time don't overrule performance during design, and that poor workmanship and contractors' substitutions of materials or components during the construction phase don't similarly prejudice the performance of the building. There is a danger that the brief focuses too much on the product (a building) and not enough on the aims (what is wanted from the building by the people who own it, use it and the social and environmental performance required of it). There is also a danger that the briefing process is governed by the market and the norms set by the supply side of the industry and not enough by the expectations of the demand side; we come back to the role of building users.

Good buildings that are designed to enhance users' experience and reduce risks to the environment need not cost more than buildings that do not. They are just better buildings. The issue is doing more with less resources, energy and waste, which is principally about better understanding. It is necessary to design in performance from the very outset; as we say elsewhere, it's no good bolting on measures to a flawed initial concept with the wrong priorities. Nevertheless, there may be additional initial investment required when one reduces the overall lifetime operational costs.

Owner occupiers, both residential and commercial, have an incentive to design for lower operational costs; renters do not and are often left with unaffordable ongoing costs for utilities and defective maintenance by both public and private landlords. In the UK, buildings are often plagued by defects on completion which, among other things, lead to increased running costs. Occupiers are often underprepared for operating building systems efficiently when faced with an array of digital devices to control overcomplicated heating, ventilation and solar hot water installations.

In the public sector, the Private Finance Initiative was introduced in 1992 and expanded by Labour administrations between 1997 and 2010. Contractors bid for contracts to finance, design, build, manage and maintain education, health and residential buildings, typically for 25 years. This pushed public expenditure into the future and was intended to give operators an incentive to provide efficient, low running cost facilities. In practice, it was often a very asymmetric arrangement negotiated by a large contractor's team of professionals, including lawyers dealing with a young, inexperienced local government official. This has led to some poorly designed and badly built facilities with large profits from 'unforeseen' costs and stiff penalties for necessary changes.

Meanwhile, government policy on promoting investment in carbon reduction has been and remains inconsistent, whether it be renewable offshore, onshore or individual power generation, individual heating systems and home insulation, with a succession of failed Green Deals and other subsidy programmes.

Value engineering

The quality of everyday items has improved considerably while costs have come down. Outsourcing manufacture to countries with currently low wages has a large impact on profits but cannot explain the orders of magnitude reduction in cost coupled with higher quality and functionality. The real reason is value engineering (VE).

Sure, handcrafted furniture will always have a quality that cannot be matched by factory-made items but the cost of new handmade furniture is beyond the reach of most people. Manufactured furniture is so much better value, whether new or vintage.

The architect Lloyd Alter was reminiscing nostalgically about the loss of the retailer Radioshack in North America (or Maplin in the UK). These stores were crammed with all sorts of electronic gadgets that are now redundant because the cheapest smartphone has higher functionality for a fraction of the cost and bulk. We have not seen the same technical progress in buildings; indeed, in many cases performance has gone down.

There was a time when the cost benefit of seat belts in cars was hotly debated. Adding seat belts has a cost, so unless consumers are convinced that they are a benefit any manufacturer that added them as standard was at a commercial disadvantage. Once they became law, the playing field was level. It was the same with engine emission controls. As these became a requirement, all manufacturers had to adopt them. With mass-produced products such as cars it is more difficult to game the system than with one-off buildings. But, as the Dieselgate scandal showed, manufacturers will still try.

As we have seen, in the UK at least, land prices are coupled to build cost. If build costs in the industry go up or down then land prices tend to adjust depending on what the building can be sold for. If higher quality becomes a requirement for new build and so build costs go up, then land prices will come down, unless demand outstrips supply. This means that the price of a plot of land is based on the potential sale price of what can be built on it minus the lowest market cost to build that building. If our driver is profit, we are forced to build as cheaply as possible to the minimum standard we can get away with.

Some housebuilders argue that higher quality standards for homes will increase prices and limit supply, but in reality any increased costs will be absorbed through adjustments to land values as has happened with previous advances in regulations.[1]

If you are reading this book, we can assume you want to build even better buildings, beyond code minimum standards. Perhaps this is for purely altruistic reasons or perhaps you are a developer and see a market advantage. However, recent government studies, the Letwin Review of Build Out Rates (2018)[2] and *Tackling the Under-supply of Housing in England* (2023)[3], did not find higher standards to be a constraint on the supply of housing. If higher standards cost more - as we might reasonably assume they will - then we either need to subsidise the build, adjust the land value or sell at a premium. These are valid options, but there is another way: genuine value engineering.

VE is based on a methodology developed out of necessity during the Second World War by Lawrence Miles, who worked for the General Electric Company in the USA. The aim of VE is to improve function for no extra cost or to reduce cost without any loss of function, *i.e.* to maximise value. Function can include anything we can define, although 'good design' and 'delight' are harder to measure than heating demand or usable floor area.

> **Value = (function + quality + performance) / (cost + time)**

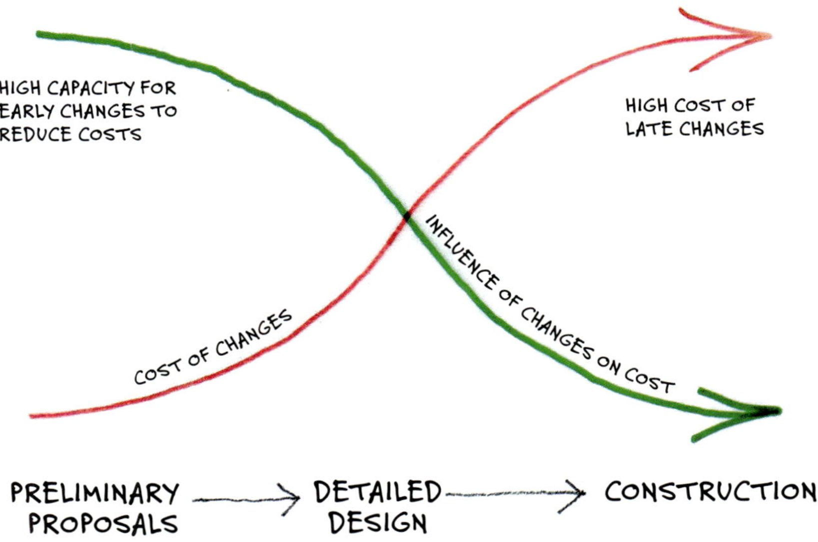

RIGHT A version of the classic cost influence graph.

There are techniques to facilitate VE, but like any creative process there is much work and no guarantee of success, with inspiration and chance also playing a role.

Most costs are incurred at a very early stage in design, before planning has been obtained and this is where the largest savings can be made for the least effort, as illustrated.

Barriers to VE in construction

As a design engineer who studied VE in the 1980s, Nick was shocked to discover that genuine VE is rarely practiced in building. Instead, the term is used in a derogatory way to describe cost cutting once the completed design has been found to be over budget. Cost cutting is the antithesis of VE and usually results in reduced value. Thus architects despise the term with passion.

If VE really can deliver more for less, why is there such resistance applying it to buildings? In manufacturing, the implementer of VE reaps the benefits and gains a commercial edge, but in construction there is often a conflict of interest between the designer and the builder. The design team want to maximise function, including wow factor, perhaps to win an award, but have little incentive to reduce cost. Budget considerations are usually limited to a target floor area and an assumed build cost per square metre.

The builder usually comes along later to bid for the work, largely on price, so is concerned about not losing money and ideally making a profit. Because of this, an experienced contractor will spot costs such as additional steel work but will also recognise patterns from previous projects as 'tricky details' where unexpected costs were incurred. Typically, this results in the sucking in of breath and the phrase 'not cheap', but unless the builder is part of the design team from the start, their input will be limited to suggesting cheaper, and probably inferior, materials and components - the dreaded cost cutting.

One way out of this bind is for designers and builders to work as a team, ideally on repeat projects so that lessons from the build (and use) feed back into the design. The examples below were developed by an integrated team of client, architect, energy consultant, structural engineer, services engineer and builder working on repeat projects, with additional value-adding input from the window and ventilation supplier. Unfortunately design and build (D&B) has an even worse image than VE!

Simplicity of form - embrace the box

Any deviation from a rectangular plan will increase costs per square metre of useful floor area - this is arithmetic. We can do the sums, but square metre build rates do not include factors for additional corners beyond the standard four. The extra building fabric costs more but also increases the heat loss area: the external walls, floor and roof. We looked at the form factor in Chapter 2.4 (heat loss area divided by TFA), and it gives us a really good indication of the relative cost, upfront carbon and heating energy demand of a building.

Passivhaus advocates are keen to point out that Passivhaus does not need to be a box, but if we are serious about delivering Passivhaus levels of performance for all, we need to *think inside the box* and stop apologising about houses that look like houses.

ABOVE Span housing, Blackheath, by Eric Lyons and Geoffrey Townsend 1964. Modest and timeless, built 60 years ago and remaining very desirable. However, improvements that might change the look are generally not allowed.

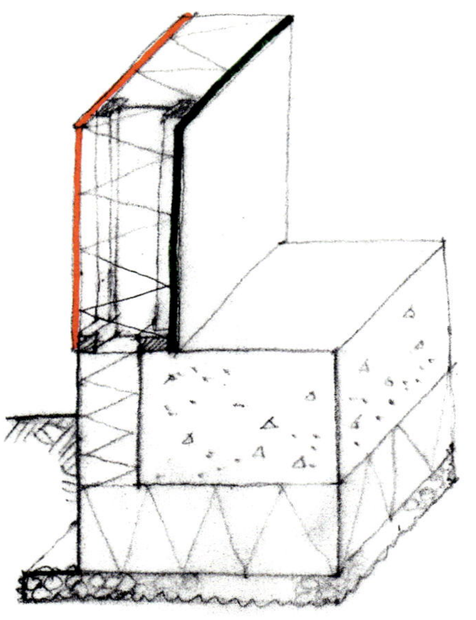

Fenestration

Windows and doors cost between five and ten times as much as the wall area they replace but they add considerable value in terms of daylight, views, means of escape, ventilation and winter solar gains. However, too much glazing or glazing in the wrong place reduces value and causes overheating in summer, extra heat loss in winter, reduced privacy, less space for storage and furniture, and more glass to clean.

A direct swap from double- to triple-glazed high-performance windows might add a couple of per cent to the whole build cost but it is the form of fenestration that has a far bigger impact on costs. Excess glazing will incur costs for additional shading devices. Large and corner windows will introduce the need for extra structural support, which may include beams, steel cantilevers, wind posts and stronger foundations to accommodate higher point loads or bending moments in structural slabs. This may in turn add programme delays, thermal bridges and more complicated airtightness. Slate House on page 144 is an example of optimised fenestration to maximise value and quality.

Structural simplicity

Faced with a gravity-defying design, most structural engineers assume they cannot question the designer and will embrace the challenge of solving the problem through ingenious (and expensive) structures and components. An experienced builder will have a deep-learned intuition for where hidden structural costs or savings lie, but this experience is rarely present in the design team until it is too late to design out the problem.

The margin illustration shows the wall slab junction we use on most of our timber-frame houses. The balloon frame 'I' stud walls are supported on the slab edge. With no manufacturer data available, simple testing was required to prove that the web was more than capable of carrying the weight of cladding. Earlier iterations were much less elegant with a separate structural frame and custom-made Larsen trusses. They cost more, took more time and were less thermally efficient. While difficult to see what could be removed, we are already considering simplification.

Embrace airtightness and reduce insulation

The main concern for a first-time Passivhaus builder will be meeting a contractually binding airtightness target 10 times more onerous than they already may struggle to meet. Conversely, experienced teams realise that with simple details and a bit of care, they can consistently achieve air leakage less than $0.2m^3/(m^2h)$ at no extra cost – the definition of value. These teams see airtightness as a saving, with less insulation needed if they can guarantee an n50 under $0.3h^{-1}$ rather than the Passivhaus minimum requirement of $0.6h^{-1}$.

While some designers take pride in the large number of details they provide, we should be aiming to eliminate the need for any airtightness drawings by making the strategy simple. A recent self-build home with inexperienced builders required only four A3 drawings to detail all the air and weathertight junctions. We made one visit to demonstrate taping and the first blower door result was $0.09\ m^3/(m^2h)$. Simplifying airtightness details tends to simplify construction in general.

The photograph below is a refinement of an old idea: sprocket rafters. The builder, Mike Whitfield, used the same detail on the uninsulated garage because it was quicker and more robust, even though the wind-tightness was not required. Purlins and ridge beams are avoided if possible, saving cost and crane hire. If they are necessary, they do not penetrate the air- or wind-tight layers.

LEFT Sprocket rafters avoid rafters penetrating the weathertight layer to create an overhang. Quicker, cheaper, better.

Passivhaus and VE

This next section will mean more if you are familiar with the Passivhaus approach (see Chapter 2.6), but the principles apply to any energy, carbon and comfort targets.

Passivhaus is often seen as an aspirational standard but it is well suited to supporting true VE. To achieve this, the Passivhaus approach, including modelling in the PHPP, must be followed from the start of the project. However, at this point fees are rarely available for specialist consultancy but it is early decisions about form, wall thickness, fenestration and orientation where most value is locked in. A number of professionals do not offer Passivhaus as an extra, but like seat belts in cars it is included as standard. Once Passivhaus, or any equivalent performance target, becomes a constraint rather than an aspiration, a way will be found to achieve it.

Key to this is embracing the necessary modelling at an early stage, but to do this it will need to be cost neutral. How is that possible? One way is to base the design on previous projects. Rules of thumb and a pattern language for genuinely low-energy buildings could allow a cost-effective Passivhaus and low-carbon vernacular to evolve. Even without any energy modelling pre-planning, we should not be far off target.

In terms of actual modelling, the smaller the budget, the easier it is to achieve this for low cost. Indeed, the model becomes a tool that helps inform decisions and can speed up our design process. For a simple rectangular house, the heat demand and summer comfort can be modelled in the Passivhaus Energy spreadsheet in half an hour using simple hacks. Dimensional inputs are limited to number of floors, internal width and length, ceiling heights, floor depths and roof pitch. External dimensions are then derived from standard build-ups in the U-values sheet. Our standard thermal bridge (psi) values are already in the spreadsheet from the previous project and the lengths are read from the dimensions already calculated. An estimate of internal wall length and width and stair dimensions allows the TFA to be estimated. For a small house this is the most sensitive variable, and uncertainty about stair and internal walls adds an unnecessary level of risk and complication at the design stage, even though they don't affect fuel bills or comfort.

Windows and installation psi values are in place from the previous project and can easily be tweaked for site-specific views and daylight. Adding a cell in the area sheet allows us to rotate the building and our standard reveals are already in the shading sheet but easily changed. We have immediate feedback and the PHPP can track the design as it progresses. Certifiers advise designers new to Passivhaus to allow plenty of slack in the PHPP at an early stage, but this can add considerable cost that is hard to remove later.

Another potentially costly aspect of any consultancy fee is site visits, especially if the project is some distance away. For early projects, we would visit the site regularly to solve problems that were often airtightness related. Experience shows that details that 'will be easier to sort on site' are never easy. For a recent self-build project site, visits were offered as an extra. Because the house went up so easily, the client did not feel the need to pay for a site visit. Keen to learn from the build, we visited anyway on the pretext of demonstrating how to install a window. We arrived in heavy rain so proceeded to install a window behind

 Tilt and turn windows allow external blinds. These provide solar control and insect screening for summer ventilation.

the weatherproof wrap. The rain continued for weeks so all windows were installed this way, with the external membrane only taped to the windows once the weather had improved. Had it not been raining as we arrived, we would have cut a hole in the wrap before installing the window. Having been prepared to write the day off as research and development, the client was happy to pay for our time and we all learnt a lot.

Zero site visits are least cost but unlikely to be best value for the client or design team needing to learn. Airtightness testing, using a fan mounted in an external doorway or window to pressurise the building, is particularly valuable in terms of providing feedback on what works.

Conclusions

While cost cutting is done by haggling and shopping around, VE is a creative process with uncertain outcomes. Often the simplest solution takes the longest to achieve. We do not claim credit for any of the examples shared in this section; all are the result of input by many people and some are simply adapted from standard details. Many of these innovations happened solving problems on site so the saving was only realised on the next project. Savings only happen if the budget is reduced, otherwise time and costs expand to fit; building is never under budget!

The only thing most architects hate more than VE is D&B. The marriage of these could be rebranded as integrated design. Imagine a team that includes the client, end users, designers, engineers, product suppliers and contractors all working together for a common goal. Something needs to change fast if sustainability is to become normal, which is part of the very definition.

ABOVE This prototype prefabricated window buck simplifies weathertight detailing and would be ideal for off-site manufacturing.

Appointing a team

Finding the right consultants and contractors can be a real problem; not too big a firm to get a personal, attentive service yet not too small to be short on resources, breadth of expertise and experience. The cost of professional consultants can be significant, especially for individuals and community groups with limited financial resources. Modern development is complex and relies on inputs from a range of experts, which may include all or some of the following:

- a project manager;
- a design team, including architects, structural and environmental engineers and cost consultants;
- planning consultants to navigate the planning system;
- fire experts and approved inspectors to comply with building control;
- lawyers, financial and mortgage advisers to deal with land and finance;
- an employer's agent and a clerk of works to oversee construction;
- a management agent to deal with landlord matters;
- possibly a management contractor;
- and a client's representative.

Finding the right design team

While most carry out their role with the utmost professionalism, putting their clients' interests first in their dealings with state and commercial bodies, there is nevertheless a persistent, not unfounded distrust among some members of the public that this may not always be so and that 'experts' may act in a self-serving manner. Walter Segal had a very robust attitude towards his relationships with his clients and wrote, 'I never use my clients and their resources as a means of self-expression. Solutions should be optimal and typical, not individual. A result thus obtained is rarely startling. The better a design performs, the less pronounced it will appear, the less assertive.'[1] Oh, that some designers paid heed to this condemnation of assertiveness.

Segal also had things to say about another area of distrust of experts – the place of non-experts in decision making:

> *We could not only bring back the joy of building but thereby make people happier. But this can only be done if we succeed in involving the public to a much greater degree.… It will give us [as architects] a new status, because we will no longer be people who are on the defensive. Then we will be able to offer so much more than now. There is a great deal of design ability and design desire in the public, and this must be released.*

Not all – or some might say not many – consultants think this way. Jon's view is that it makes the job of design so much easier and more effective if you have a brief articulated by the users of the buildings and are not trying to conjure up out of thin air what users of the building might think or feel about the design. And not just designers; managers and policy makers would also benefit from working with the users of their services. Again, Segal had a good word on this: 'The technician's job is to clarify the technical possibilities so that the people's demands can become more precise and lead to action.'

Cost consultants

One particular requirement generally unmet is the need for reliable information on the cost of construction. For some reason, experienced and well-paid quantity surveyors find it beyond their powers to predict costs or to advise at an early stage as to where savings could be made or expensive mistakes avoided. Many is the time that projects have to be 'value engineered' when the tenders come in.

Quantity surveyors seem reluctant to engage in a critical appraisal of costs at an early stage, although they have extensive experience from many projects of how cost overruns occur. Is this because the consultant QS is culturally and educationally discouraged from taking part in discussions about building design? They should be able to predict where cost overruns are likely from the design or other building practices.

The role of enabler

It's not just cost. In housing development, you need good advice on a wide range of
matters: how to obtain land, loans and grants as well as organising groups and
understanding the local politics. In parts of continental Europe where communities are
more engaged in co-housing and co-operative housing of various forms – in Germany,
Switzerland and the Netherlands, for example – a consultant role has emerged of 'enabler':
one who is knowledgeable and experienced in that range of areas. They often come from
architectural, planning, housing or development backgrounds and there is an initiative by
Community Led Homes, who have established a network of consultants with this range of
skills in the social sector in the UK.

LEFT Not just design
participation but a two-way
learning process with
community members dealing
with finance, regulation, politics
and legal matters.

Competitions

Also in Northern Europe, and Scandinavia in particular, many or most large or not so large
projects are awarded by competition. Competitions have a poor reputation in the UK for
producing unworkable designs and excessive budgets. Competitors complain that the
requirements are too onerous in terms of time and money. However, some competitions
are limited in the UK to a selection of participants invited for their particular relevant
experience or expertise for which they are paid a modest fixed fee.

What is important is to have a clear brief that sets out what is wanted of the proposals
and does not use the word 'innovative'. Vital is a realistic budget with a mandatory
requirement that competitors demonstrate how their submission meets the budget. The
judges must fully understand the clients' requirements. Sponsors have a duty to inform
unsuccessful competitors why they were unsuccessful. This allows them to gauge how they
stand with their competition and they can feed back into subsequent submissions.
Unfortunately, in the UK many sponsors do not recognise this duty, making entering
competitions doubly unrewarding. The rules governing competitive tendering for building
contracts make this duty mandatory.

Procurement of a contractor

To create a good building you need a good contractor: careful, trustworthy, well-organised, knowledgeable and experienced. You also need good advice; consultants with similar characteristics, which we have discussed earlier in the chapter. So, the question is how to find them, agree what they are to do, work out how much to charge and decide how to resolve issues if things go wrong.

Generally speaking, it is difficult to legislate for the qualities that are required; it is a question of attitude with a degree of experience. Recommendation counts for a lot – from former clients or fellow professionals. One can learn much about the context for a project by talking to a range of people. It is important to find a match with the size and type of project. Take up references and interview candidates, and if you are looking for a builder, make sure you talk to the people who will be in charge on site.

It can be of great advantage if the consultants and contractors have worked together on previous projects. They will be familiar with their respective aptitudes and will have had opportunities to develop designs and approaches that work well on site. This led Mark Brinkley, author of *Housebuilder's Bible,*[2] to suggest that for small projects, rather than appointing an architect in the first instance, there is merit in approaching local builders and asking them who they would recommend.

Establishing the price can be difficult. The default tends to be some form of competitive bidding process. The requirement should be clearly defined, otherwise it can be difficult to compare offers. Nevertheless, bidders will often make their own assumptions about what is required, which can sometimes be useful, as you can learn from other people's ideas of what is required, or at other times can just cause complication. There are rules designed to prevent bidders spending time and effort preparing a bid being unfairly treated by not winning, even though they have made the cheapest offer. However, there is a suggestion that it is unwise to accept the lowest bid because there is a risk that it is too low and will lead to challenge throughout the contract to recover costs.

While you may not be able to legislate for quality and performance, you do need some mechanism to define what happens when things go wrong and work is inadequate, late or over cost. If the attitude of the parties involved is positive, it makes the significance of the various forms of contract less critical. Nevertheless, the different forms change the balance of responsibilities between the parties.

Design and build

A so-called 'traditional' contract is based on a relatively complete and detailed set of information: drawings and specifications of the building, to be constructed for a stated 'contract sum' within a stated 'contract period'. This form of contract can offer the client, or 'employer', the ability to amend, add or omit sections of work and the contract sum and contract period are then adjusted following a negotiation based on rules intended to arrive at a fair, amended cost and programme.

Difficulties often arise with controlling costs because the contractor is in a strong position when negotiating a price for any work that has been varied. This contract also offers the contractor the ability to claim for costs arising from eventualities not under the control of the employer, such as delays caused by inclement weather. However, such a contract is appropriate if you are particularly concerned about quality or if sections of work are likely to be varied, added or omitted, which may be the case in a project where self-builders are proposing to carry out part of the work.

A significant limitation of the traditional approach is that the contractor is unable to contribute their experience and expertise to the process of developing the design to make sure that it is not complicated to build and therefore more expensive than necessary. D&B contracts make the contractor responsible for developing the design at an earlier stage. This is generally after the design team has developed a proposal to concept design stage and obtained planning permission and before the detailed technical design is undertaken. At this point, the contractor would be employed – following a competitive tendering process that fixes the cost and contract period – to develop the design according to a set of 'employers requirements' and go on to construct the building. The contractor employs the design team and is responsible for detailed design and specification.

The principal advantage is said to be certainty of cost and time. However, the cost will tend to be higher because the contractor is carrying the risk of increased costs and delay during construction, and clients often disregard this. Also, in practice, circumstances sometimes arise forcing a variation to the work, which carries an increase in time and/or cost.

The role of an 'employer's agent', who represents the client once the contractor assumes design responsibility, is critical. They often do not have the technical knowledge or experience to match that of a design team and they also tend to have a very hands-off approach to monitoring work on site and checking quality.

The UK construction industry has moved from general contractors employing tradespeople proficient in the various skills involved – bricklaying, plastering and so on – to outsourcing most of the actual construction work to specialist subcontractors. The success of a D&B process depends to a large extent on the attitude and competence of the subcontractors who are operating in a very competitive market, working to a fixed price and timescale. This drives a race to the bottom and there is a danger that they will skimp on the work and/or substitute cheaper materials and components without informing the main contractor or employer's agent. This cut-throat approach to subcontracting is a scourge on the UK building industry.

ABOVE Flats by Jon at Architype: a simple block with bolt on steel stairs and balconies. Very straightforward construction of blockwork with external wall insulation and pre-cast concrete plank floors developed in close collaboration with the Design and Build builder/developer client. D&B as it should be.

Russell Curtis, a director at RCKa architects, was prompted to comment in *The Architects' Journal* in response to the Grenfell Inquiry:[3]

Design and build is construction's dirty little secret, a seductive contractual brew which purports to offer clients certainty and comfort …. D&B is not, in and of itself, the problem. Rather it's the manifestation of a rotten culture of risk-avoidance, which has demoted the pursuit of quality so far down the list of priorities that in some cases it ceases to factor at all. Yet, it is easy to understand why it remains so popular with clients. The opportunity to offload the risk of cost and programme overruns to a main contractor, thus saving face when things go awry, is too tempting to ignore … In response, contractors – working to razor-thin margins – simply bat the risk down the supply chain to the design team, suppliers, specialist subcontractors, or whoever happens to be in the room at the time.

Nevertheless, D&B when practised as a properly integrated design process is potentially the most effective procurement route. Done badly it creates poorly designed and constructed buildings at an uncompetitive cost. Success requires the contractor to appoint and work closely with the initial design architect to carry out the detailed development of the design and to have a balanced view of quality and cost, and for the designer to respect the skill of the contractor in executing the work.

Nick and Jon have experienced the challenges of D&B from both sides. Sometimes they have handed over what they considered to be a well-considered design only to see it

diluted by cost cutting. At other times they have worked with the contractor as part of the design team to help deliver a good building that went to D&B contract with serious design defects which can be hard to fix post-planning.

Other ways of involving the contractor in the design

There are a number of variations of this theme designed to reap the benefits of involving the contractor early on in the design and construction process while introducing a level of competition to ensure that costs are reasonable and contained. This may involve employing a contractor at the outset and paying them for their time taking part in the development of the design using a Pre-Construction Services Agreement (PCSA). This would be followed by a second-stage tender for carrying out the construction work. This contract is often awarded to the same contractor that carried out the initial design work because they are familiar with the detail of the design.

Other variations of the early involvement of contractors include *construction management*, where the client employs a consultant construction manager appointed by the client. This procedure often includes involving the various trades contractors in the development of the design. A similar arrangement with a confusingly similar name, *management contracting* is where a client employs a management contractor to co-ordinate the design development and construction by a number of trades contractors, but this time employed by the management contractor rather than the client. Both construction management and management contracting enable the experience and expertise to be brought to bear on the design from early on. Management contracting is simpler for the client as they only have to deal with a single contract rather than all the separate trades contracts which are administered by the management contractor.

Our experience with engaging contractors early in the design process is patchy; some contractors with long-standing relationships with clients and designers embrace this approach fully, whereas others are happy to take the fees but do not engage fully and bring little useful technical or cost information to the table.

There are other approaches to obtaining an efficient and cost-effective process. 'Open book' contracting is where the builder is fully reimbursed on the basis of transparent records of the costs they have incurred. The expectation is that you can save money because the contractor does not have to price the risk of costs rising for whatever reason. There is a risk that the contractor is not incentivised to secure good value for money when negotiating subcontract packages and supplies.

A variation on this arrangement is to establish a 'target cost'. The client and contractor share any gain by notional savings of actual costs against the target, or any pain through notional losses. The advantage of a target-cost arrangement is that it incentivises the contractor to build within the budget and achieve savings whenever possible.

Self-build

For small-scale self-build projects, doing all or some of the work yourself can achieve significant cost reductions. This view is not shared by builders who have worked with self-builders who have failed to understand the skill, knowledge and attention to detail required for some apparently simple tasks. This may not involve carrying out any of the actual construction work but merely taking over the main contractor's role and organising a series of subcontracts for the construction. This will save the contractors profit, which may be up to around 15% of the cost, but be warned: many people underestimate the amount of work and knowledge involved. The job is made more difficult because an established contractor has a network of contacts to form a building team who are familiar with working together in an efficient manner.

At the other end of the scale, you could do most of the construction work yourself and save up to half of the total building cost. Again, savings may be reduced because an established contractor has a network of reliable suppliers who will offer discounts for quantity and repeat orders. You also may well have to buy in certain elements of work that require particular skills or equipment, depending on how easy to build the design is.

RIGHT An early example of a self-built property from Ipswich, 1955, at a time when groups of tradesmen (and they were almost always men) pooled their skills to build groups of conventional homes.

The contractor was involved on site, but at a cost

The couple who commissioned this large house designed by Jon and taken to completion by MAP Architecture were deeply involved in all the decisions affecting its look and performance. They worked with the contractor on site to achieve what they wanted but this increased the cost because they employed a general contractor rather than employing and managing sub-contractors directly and because the brief and some of the details were complicated.

LEFT AND BELOW It required a good deal of ingenuity and hidden structure to achieve a simple looking solution. We touch on the cost of architectural minimalism in Chapter 2.3.

The curved grass roof is designed to 'float' above the building.

Risk management and regulation

Avoiding unnecessary risks is a laudable aim, whether of death, serious injury or budget overrun. Building sites have become much safer since concern was raised and legislation introduced. One might imagine that safety in fire would be prioritised as a given. However, the inquiry into the Grenfell Tower has been marked by all the organisations involved in the design and construction and regulation of the refurbishment. Responsibilities were poorly defined and split between different parties and this shows how not taking responsibility can have deadly consequences. As is now widely recognised, Grenfell is far from being an exception.

There are risks involved in any activity. The important thing is to be aware of them, how likely they are to occur and how serious it may be when they do take place. Consider how different risks can be mitigated; how you would take steps to reduce the risk of a project running over budget – for example, by carrying out regular cost checks, mobilising more resources to stop the project running over time, and so on. Costs can be put against the various risks, and contingencies put in place to cover them. This means you can have a measure of the impact of the eventualities that may occur. Some risks are existential and beyond control; others can be dismissed with a phone call. The important thing is to be aware of what may come to pass and to have an idea of how you would react to minimise any negative effects.

Designer risk assessments are supposed to help design out risks (or put in place measures to manage risks that could not be designed out). If embraced at an early stage, this can be a really useful process but often it feels more like an administrative chore, like filling in your tax return, and it is left too late.

While it is good to design out unnecessary risks, there has become a tendency to imagine risks where few really exist, leading to time lost and opportunities for progress missed. There is a real chance that ambitions are reduced to avoid some imagined risk. A risk-averse approach tends to avoid addressing issues – not challenging decisions by officials looking for the easy life, for example. This attitude reduces ambition to the lowest common denominator; for minimum trouble read minimum effort. The balance between challenge and accepting the status quo demands judgement but a culture of avoiding risk rather than recognising problems and solving them leads to a reduction in aspiration, which is not conducive to progress.

Risk and responsibility

There is an interesting overlap with risk and responsibility. Doing things 'by the book', even when you suspect the rules are flawed, is a way to avoid being held responsible. However, as professionals or just responsible human beings, we have an ethical responsibility to flag up problems we see or propose better ways, even if these are not the norm and so may be seen as incurring risk.

Similarly, when things go wrong, which they surely will, professionals are advised by their insurers not to admit liability. This can be painfully frustrating and delays the resolution of problems.

In the world of community-led development, additional challenges are present; you are talking about organisations without experience, assets or particular expertise taking responsibility for the relatively complex business of development. People coming to it for the first time are generally surprised at how complicated it can be to find a site, raise finance, buy the land, design the project, obtain the necessary permissions for how to plan and construct it, sell or let it and manage and maintain it. There are additional

RIGHT Narrow streets for pedestrians, bikes and cars in Vauban, Freiburg.

responsibilities if you take on the construction of all or part of the development.

This is all made doubly difficult in the UK because there is not a great deal of recent experience or governmental, financial or professional infrastructure to support this approach to development. Instead, we have a culture of mutual mistrust by and of professionals, officials, contractors and financiers who do not understand what ordinary people are capable of organising and building. They often tend to see such enterprise as a hugely risky venture. This is in direct contrast to experience in Northern Europe where governments encourage people to get involved for the social benefits that follow, where there is a profession of enabler (see page 255) who manages community-led projects dealing with local authorities, designers and financiers and where lenders see community groups as a good risk; they cannot afford to fail – and don't.

In Germany and elsewhere on the continent, regional banks have direct relationships with the local authorities and housing professionals in their area and record very low defaults on loans. It has been reported, for example, that in the new neighbourhood of Vauban in Freiburg, in which 2,000 low-energy apartments have been developed by 40 community groups, only one group had to defer repayment of their loan by one month due to cash-flow problems over a 10-year period. This is in contrast to banks in the UK who prefer high returns on global financial gambling. Social lending in the UK is also constrained because social lenders are not investing in enterprises with high returns and are nervous about risky lending.

Regulations

The design and construction of buildings is regulated in many ways. Planning law controls land use and the overall size and design of buildings. Building control governs fire safety but also structural safety, energy performance, drainage and other aspects of the performance of buildings. Funders require structural insurances and fire precaution measures to limit losses caused by fire. Designers have to ensure that buildings are safe to maintain and contractors must conform to health and safety legislation during construction.

You might imagine that safety in fire was a given; something beyond compromise. However, the tragedy at Grenfell Tower, where 72 people died in a fire that spread quickly through the 24-storey building, reminds us of the need for regulation to uphold standards of safety. The inquiry into the disaster has revealed a culture of cost cutting, negligence, dishonesty and disregard for the residents among designers, manufacturers, suppliers, contractors, government and within the local authority as regulator and client. Materials testing was falsified, evidence of risks of fire disregarded, regulations compromised and not enforced, and the expressed concerns of residents ignored. While not all of this behaviour was criminal, much of it showed a startling lack of duty of care towards those who lost their life. This episode had extreme repercussions but does suggest that there may be a more general lack of care within the industry and society in general.

Building regulations

Hopefully regulation can prevent the worst corruption within the complex system. However, what it is unable to achieve on its own is the care and commitment to doing the right thing necessary to produce good buildings – buildings that serve their users as well as they can. This requires clients, designers, suppliers and contractors to go beyond the minimum standard set by regulation, which can be progressively raised to improve performance overall. On the other hand, the minimum standard can become the norm and serve to suppress improvements in performance.

Much of the interpretation, implementation and inspection of building regulation has been outsourced to private companies in England and Wales (but not Scotland or Northern Ireland where it remains a local authority duty). The outsourcing of professional advice and research from the civil service by the government has led to potential conflicts of interest and suspicions of vested interests distorting the outcome of studies and test results. Little research into the performance gap between the planned and actual performance of buildings, for example, is carried out by the private sector.

The high cost of product testing and certification has also become a significant barrier to progress. There are suspicions surrounding the quality and impartiality of advice under the building regulations as companies compete by offering favourable interpretations of the regulations for the minimum fee.

This competitive market in services also appears to have reduced the amount of monitoring on site. Jon once witnessed a visit from a building inspector employed by a company of approved inspectors who walked into the site office, where he was handed the site inspection logbook by the contractor and he ticked off the boxes that all had been inspected and passed as adequate. Jon then went outside to look at the work himself to discover that the drains for the housing estate had been laid so that they emerged out of the ground ending up over a foot in the air and well above the ground-floor level of the houses!

Structural insurance

Before you can obtain a mortgage on a new property, lenders require insurance against structural defects in addition to any building control consent. However, such insurance does not generally cover the first two years, when most defects appear, or after ten years when cumulative problems become apparent. The cost of this insurance is significant and, of course, passed on to the buyer who little realises that such cover is of limited value and is more of a marketing ploy than anything else. It will also usually only cover 'conventional' construction and acts as a brake on the development of improved ways of building.

Health and safety

Jon's first experience in the building industry in the mid-1960s was as a labourer. One of his first tasks was to hold the one good leg and other peg leg of a bricklayer (who had lost his leg when a steel beam fell on him) as he leaned over the parapet of a 10-storey block of flats to do some pointing. Jon was terrified that he might not be able to hold him and that they would both plunge 100 feet to their deaths. The brickie seemed unconcerned.

Things had changed a little by the time the Lewisham Self-Builders came to build their houses in the late 1970s and early 80s. The Health and Safety at Work Act had been passed in 1974, but it was another 20 years before regulation of the building industry was enacted and health and safety was not high on anybody's agenda. Therefore self-builders with limited building knowledge or experience, and with very little in the way of training beyond an understanding that building is potentially dangerous and that you had to be careful and sensible, threw themselves into building their own homes from the ground up. Scaffolding was not used and families with children were encouraged to be on site. There was not a single accident or injury as far as Jon is aware – then and over the next decade building another 200 or so self-built homes. Meanwhile, as health and safety became embedded as a necessary concern, this approach is no longer possible; it is now hard to imagine running a building project without scaffolding and with children on site.

Since 1994, health and safety has been governed by the Construction (Design and Management) Regulations, which imposed duties on clients to deal with safety. These ensure that designers and contractors have the relevant knowledge and experience to manage safety on the project: on designers to eliminate or reduce foreseeable risks, on contractors to plan and monitor safety during construction and on workers to cooperate to take care of safety. These duties involve preparing assessments of risks, a construction phase health and safety plan together with establishing a health and safety file for use during the project, to be handed to the building owner on completion at design stage. This imposed a significant bureaucratic burden on projects and spawned a whole new area of specialist health and safety consultancy and new role of CDM co-ordinator to orchestrate the assembly of the necessary paperwork at successive stages of the project. In an attempt to simplify the process, the pre-contract coordinating role has now been incorporated into the duties of the principle designer.

The introduction of health and safety at work legislation has significantly reduced deaths and injuries and has encouraged the industry to take safety more seriously. It requires designers to design out risks during construction and use as far as possible, but hazards inevitably remain and legislation requires those involved to clarify remaining risks and assume responsibility for taking action to avoid them. The bureaucratic burden remains and calls into question whether wrapping the issue up in reports, plans and risk registers ignores people's innate propensity to avoid risk.

The Grenfell tragedy has prompted the Building Safety Act and the Fire Safety Act to tighten up the fire safety of 'high-risk' residential buildings by introducing more stringent duties on designers, contractors and manufacturers and new regulation of professional competence requiring training, proof of competence and regulation of product information. Also required is a 'golden thread' of digital safety information to be kept up to date with any variations during the life of the building – all of which address the problem with additional oversight. What it does not do is establish a clear 'golden thread' of responsibility.

A contrasting approach is offered by Aldo van Eyck's Amsterdam playgrounds, 700 or more laid out between the 1940s and 80s. He designed them using simple elements that invited children to develop an awareness of potential danger, rather than protecting and isolating them. The playgrounds were unfenced and many were adjacent to busy roads. Kids co-operated by looking out for cars and developing rules for play that avoided being endangered by traffic, while adults occupied benches strategically placed so that small children could be supervised. The self-builders in Lewisham and elsewhere had a similar approach with children on the building site, with adults around to supervise them as they played and found their way around the potential dangers – all without mishap.

Planning control

The Town and Country Planning legislation introduced after the Second World War was intended to direct the national reconstruction effort, to ensure that land uses were compatible, housing conditions were good and there was access to light, privacy, transport, schools and other services. There are layers of policy at national, regional, local, neighbourhood and even individual site levels and these are sometimes contradictory, ambiguous and unclear. Over time, more resource has been put into 'development control' rather than the strategic planning of infrastructure, density, height and land use.

Complexity in the planning system

Planning has become the detailed regulation of aesthetics, materials, details and other largely subjective matters applied to individual developments – some of which may be as small as fitting a roof window or a case recently of painting the walls of a building. Attention is concentrated on external appearance – 'materiality', whatever that is – proportion, and making sure that new buildings 'respect' (*i.e.* look similar to) existing ones. Planners are only interested in considering a design from the outside, paying little attention to what it may be like to live in the building: is there a view or sufficient sunlight and daylight and is the size and layout of rooms satisfactory?

This preoccupation with aesthetics serves little purpose; it can lead to uniformity and the lack of variety that enlivens a typical country town. It stifles progress, limits people's choices and freedom of expression, while draining resources from strategic decision making and causing frustration amongst applicants who see some of the measures unnecessarily limiting what they can do.

A recent permission for a development of 36 dwellings in two four-storey buildings came with no less than 37 conditions, most of which required extra work and significant cost to comply with. Meanwhile, planning control has not prevented a great deal of unsatisfactory development going ahead. Good buildings and environments have been lost to be replaced by inappropriate developments driven by commercial incentives.

Walter Segal was mystified and outraged in equal measure:

The most curious phenomenon of the post-war period is that together with the free for all in architectural expression there arose in the deeply divided professions concerned with the shaping of the environment, the demand for visual control in planning … in the complex community of the country, visual control by coercion is intolerable.

There is little evidence that the community has been well served by the controls it has set up in visual matters.

OPPOSITE In 2022, Oxford City
Council made John Buckley's
protest sculpture, the Headington
Shark, a heritage site for its
'special contribution' to the
community, despite objection by
Bill's son Magnus Hanson-Heine
who pointed out the absurdity of
preserving a symbol of planning
law defiance.

Uncertainty in the planning system

There is uncertainty about the basics of what you can do where, which has led to a lottery in the market for land. Developers insure against this by buying options on any parcel of land that may conceivably become developable. This increases risks and tends to limit the market to those in the know. Meanwhile, if you are lucky enough to win this lottery, you will be faced with a lengthy and complicated process to determine the detail of your proposals. This may involve commissioning costly reports on energy performance and biodiversity as well as privacy, daylight and so on. These reports are costly but often contain a lot of regurgitated planning policy copied from the last project, which is largely unread. Much of this could be avoided by requiring decent across-the-board standards for energy and so on. Planning officers often do not trust non-professionals and so a planning consultant has to be commissioned, whose job it is to commission these reports and negotiate with the planning officers. It often takes many months beyond the statutory period for resolving planning applications, which puts development programmes in jeopardy.

Subjectivity in the planning system

The process is opaque, subjective and inherently undemocratic, without adequate methods for neighbours and other interested parties to be consulted. Planning departments are under-resourced and staff are poorly trained. It is often difficult to talk to planners other than at pre-application meetings, which take time and money and often do not clarify issues and reduce risk. There is too much attention to what things look like rather than how they perform and a new language has arisen that is beyond understanding. For example, a recent statement supporting a planning application for a new housing development in London described a typical annexe as a two-storey element with a mezzanine level, whose connection with the main house is linked by its primary materiality at ground-floor level. It suggested that its separateness is subtle and announced to the street through the materiality and form used to the upper level and that delineation of the plot was signified to passers-by through a surface continuation of the brick wall materiality which forms the defensible space.

The effort to legislate for quality has not prevented bad building; better to set the framework and let people get on with it within the constraints. Such an approach is better suited to the ongoing nature of development, whereas the current emphasis on detail implies heading for a fixed 'end point' that does not exist.

Planning elsewhere in Europe tends to be more strategic and less prescriptive of detail with lower concern for appearance. Planning controls set up a basic framework governing infrastructure, land use and the density, footprint and height of buildings. Within this framework, you are able to design to as large an extent as you wish. The rules are explicit rather than based on subjective views on what things look like, and everyone knows what you can and can't do.

New legislation following the Grenfell disaster – and the Building Safety Act 2022 in particular with a regulator with significant powers – is expected to make developers, contractors, material suppliers and others in the industry take building safety and the building regulations more seriously.

With regard to planning control, The National Planning Policy Framework (NPPF) came into force in 2012 with the aim of simplifying and speeding up the planning system and providing clear direction for planning policy. It has achieved much of this agenda, but the detailed control of development still remains and is the source of frustration for overstretched planning departments and applicants alike. The most recent amendment to the NPPF in 2023 reforms housing delivery and provides for more support for CLTs – hooray.

Afterword

So... hopefully we are nearer to knowing what a good building is - or more probably what a bad building is: possibly impractical, overly preoccupied with appearance, inefficient in the use of materials and energy and too costly.

We have looked at ways buildings can be improved, not getting carried away by architectural expression and superficial style. Modernism, for example, with the abstract concept of *less is more* when the reality is very complex, difficult and expensive to achieve. Post-modernism that seeks to replace perceived elitist connotations of modernism by reaching back for popular cultural references which unfortunately do not resonate with most people's idea of how things should be. Deconstructivism which takes the component parts of a building and makes them appear to be exploding into jagged fragments . . . why?

We have suggested a balance which includes utility and performance, value and economy. A wonderful new aesthetic could emerge from this approach that may well be developed into an 'ism' to be taught in architecture schools of the future. Yet again, missing the point! Or perhaps the resulting buildings might be comfortable, energy efficient and affordable but bland and uninspiring, lacking in *soul*. The designer might choose to add some art, a splash of colour or texture or some plants. Or perhaps a more neutral building will be the perfect blank canvas for the occupants to express themselves as they wish. Perhaps not everyone is as bothered as we are about what buildings look like and that is okay. Whatever the chosen response, as we have hopefully shown, it will be far easier to improve the look than to make a conceptual building perform.

We have indicated some principles that are useful for achieving an improvement, the need for a culture of care in design and construction, involving users in the design process and taking some lessons in human scale, practical design and genuinely economical and sustainable construction from vernacular building. Lessons too from self-help and collective housing practice.

We have considered ways that buildings can be improved in the most practical and cost-effective ways to achieve their anticipated energy performance: designing for the sun, for adequate ventilation and adopting the Passivhaus standard (or equal equivalent) as a robust methodology.

We have also outlined approaches to building construction and services that are simple, effective and economical. To make this real we have provided some concrete examples but these are all works in progress, will date and may not suit your application or personal taste.

Finally we have considered necessary improvements in the wider context within which building takes place in order to create a sustainable building culture: fairness in the ownership and use of land, greater emphasis on performance and practicality in design education, honesty and responsibility in the construction industry, and the understanding that risk management and the regulation of building and planning can only ever check the worst excesses of overdevelopment, shoddy construction and wasteful use of energy. This will reduce the risk of the dangerous, awful even, but it cannot create the good. Our personal preference is for minimal regulations and maximum freedom but we know that has an unacceptable human cost.

We have no illusions that some new energy-efficient buildings will save us. What is clear is that avoiding environmental catastrophe requires a radical change in economics and social values as well as what we eat, drink and breathe; how we move around; how we share the planet with wildlife; use minerals and other resources; dispose of waste... never mind how we keep dry, secure and warm in winter and cool in summer.

That needs a commitment to a culture of the good complemented by better knowledge and understanding. As we argue through the book, we believe that good design comes from an approach and not from applying a style to a problem.

We do believe, however, that readers – architects, engineers, clients, builders or suppliers – can influence thinking on regulation that institutions and commercial organisations will generally comply with to avoid the risk of enforcement and reputational damage. This can overcome resistance from people who are fearful of increasing costs and politicians who are in fear of losing people's votes.

We opened this book with a quote from one published in 1983. The following quote is taken from the latest blog post by Graham McKay, Assistant Professor at the School of Public Architecture, Michael Graves College, Wenzhou-Kean University:

If buildings that tell lies are bad, then what kind of stories should a building be telling? – and I'm not talking about 'narrative'. The best contender I've come across is 'the story of its construction' – such as the short-lived Danish modernism. Traditional Japanese architecture and much vernacular architecture does the same. Renovating and repairing old buildings to enable their continued use is another layer of history. (Restoration is a type of lie.) I'm beginning to suspect that nothing less than the dismantling of the entire apparatus of architecture will be sufficient.

PTION VESTS
KOK SEN
國成
記
球
餐室
KOK SEN COFFEE SHOP
中山會館
北京人餐馆
BEIJING FOLK
RESTAURANT
YANTI
AUTHENTIC
NASI
PADANG
热端正宗印尼
巴东咖哩饭
CLUB
Authentic
Nasi
Padang
Yanti
Authentic Nasi Padang
Yanti
45 Keong Saik Rd
Tel: 65 6324 8268
7 ELEVEN
OPEN 24
BIG GULP

Note on the authors

Jon Broome trained as an architect but is now a designer, enabler and self-builder of two houses. He is author of *The Green Self-Build Book* and contributor to *Housing & the Environment* and *Architecture and Participation*. Jon was previously a Director of Architype, an architectural practice working on housing, education, health and community buildings with specialist expertise in low-energy design, timber-frame construction, community engagement and sustainable building.

Jon now lives in London, where he runs his own consultancy, Jon Broome Architects. Jon's longest running, but most rewarding project, RUSS, was completed in 2024 after a 15-year gestation; 36 dwellings for a Community Land Trust in South London. Jon Broome Architects devised the project, Architype carried out the design consultation and SEH Architects did the detailed design and supervised the construction.

Nick Grant is a freelance energy consultant and principal of Elemental Solutions. Although Nick calls himself an engineer, he dropped out of two universities without graduating. In hindsight he appreciates the grounding in engineering science that he found so dull and difficult at the time. He has been making things since he could hold a hammer inspired by his mum's can-do attitude. His interest in closing the performance gap between design and reality led him to the Passivhaus Standard of which he is one of the UK pioneers. Nick was a long time Trustee of the AECB and is Technical Director of the Passivhaus Trust.

Nick is an active contributor to discussions on sustainable design but is also a practical engineer and self-builder. He and his partner designed and built their own home, inspired in no small part by Jon Broome and Brian Richardson's *Self-Build Book*. Like Jon he made many mistakes,

and like Jon's house, it is a delight to live in. The builder Mike Whitfield helped for some key stages and they continue to work together a quarter of a century on.

JON BROOME

NICK GRANT

Chapter notes

1.1 WHAT IS A GOOD BUILDING?

1. Saint, A. 1983. *The Image of the Architect.* Yale University Press, New Haven and London.
2. Vitruvius Pollio, M. 1914. *Vitruvius: The Ten Books on Architecture*, translated by Morris Hicky Morgan. Harvard University Press, London.
3. Pirsig, Robert M. 1974. *Zen and the Art of Motorcycle Maintenance.* The Bodley Head, London.
4. Pirsig, R. 2022. *On Quality: An Inquiry into Excellence.* Edited by Wendy Pirisig. Mariner Books, Boston.

1.2 BEAUTY, UTILITY AND ECONOMY IN ACTION

1. Neuhart J. & Neuhart, M. 1998. *Eames Design: the work of the office of Charles and Ray Eames.* Abrams, New York.
2. Padfield, T. & Jensen, P. 1990. Low Energy Climate Control for Museum Stores. *ICOM Committee for Conservation 9th Triennial Meeting Dresden German Democratic Republic 26–31 August 1990.* ICOM Committee for Conservation, Dresden.
3. Feynman, R. P. 1986. 'Volume 2: Appendix F – Personal Observations on Reliability of Shuttle' in the *Report of the Presidential Commission on the Space Shuttle Challenger Accident.* NASA, Washington DC.

1.3 BUILDINGS AS SCULPTURE

1. Dyckhoff, T. 2017. *The Age of Spectacle: The Rise and Fall of Iconic Architecture.* Windmill Books, London.

1.4 ECO-BLING – INNOVATION AND ASPIRATION

1. Liddell, H. *Ecominimalism: The Antidote to Eco-bling.* RIBA Publishing 2013.
2. Eames, C. 2007. *100 quotes by Charles Eames.* ed. by Carla Hartman and Eames Demetrios. Eames Office, Washington.
3. Slessor, C. Cork House Review – Barking Up the Right Tree. *The Observer*, 28 July 2019. theguardian.com/artanddesign/2019/jul/28/cork-house-review-eton-stirling-prize-matthew-barnett-howland-dido-milne-oliver-wilton [accessed 21/10/24].
4. Natural Cork Council. Reports & Statistics. naturalcorkcouncil.org/reports-and-statistics [accessed 21/10/24].

1.5 / 1.6

5. Biomimicry Institute. 2024. Biomimicry Theatre. biomimicry.org/inspiration/what-is-biomimicry.
6. Clarke, A. & Grant, N. 2010. Biomass – A Burning Issue. *AECB.* https://aecb.net/download/biomass-a-burning-issue.
7. Thornton, J. Rainwater Harvesting Systems: Are They a Green Solution to Water Shortages? *Green Building Magazine*, Spring 2008. greenbuildingstore.co.uk/wp-content/uploads/rainwaterharvesting.pdf [accessed 21/10/24].

1.7 A MODERN VERNACULAR

1. Turner, J. F. C. 1976. *Housing by People.* Marion Boyars, London.
2. Ward, C. 1990. *Talking Houses.* Freedom Press, London.
3. Alexander, C. *et al.* 1977. *A Pattern Language: Towns, Buildings, Construction.* Oxford University Press, New York.
4. Monte Paulsen, Lectures
5. Grahame, A. & McKean, J. Special Issue: The Segal Method. *Architects Journal.* 5 November 2005.
6. Davies, C. 2005. *The Prefabricated Home.* Reaktion Books, London.
7. McKay, G. Modern Vernacular. *Misfits' Architecture*, 25 June 2017. misfitsarchitecture.com/2017/06/25/modern-vernacular [accessed 21/10/24].

1.8 THE MODERN VERNACULAR IN ACTION

1. United Nations, 'Our Common Future' (also known as the Brundtland Report), 1987, Oxford University Press, Oxford.
2. Borer, P. & Harris, C. 2005. *The Whole House Book: Ecological Building, Design and Materials.* Centre for Alternative Technology Publications, Machynlleth.
3. Arnstein, S. 1969. A ladder of citizen participation. *Journal of the American Planning Association*, 35 (4).
4. Turner, J. F. C & Fichter, R., eds. 1972. *Freedom to Build.* Macmillan, New York.
5. Ward, C. 1974. *Tenants Take Over.* Architectural Press, New York.
6. Atkins K. 1983. *Bulletin of Environmental Education*, October 1983. Town and Country Planning Association, London.
7. Segal W. 1982. View from a Lifetime. *Transactions of the RIBA*, V. 1, London.

1.9 SELF-BUILD

1. Ama Research, 2023. *Self Build Housing Market Report – UK 2021-2025*, 8th edn. London
2. Broome, J. 2006. *The Green Self-build Book: How to Design and Build Your Own Eco-home.* Green Books, London.

1.10 SUSTAINABLE NEIGHBOURHOODS

1. Falk, N. & Rudlin, D. 1999. *Sustainable Urban Neighbourhood: Building the 21st Century Home.* The Architectural Press, Oxford.
2. Girardet, H. From Mobilisation to Civilization. *Resurgence* 167, November/December 1994.
3. Fairbrother, N. 1974. *The Nature of Landscape Design.* The Architectural Press, London.
4. Chatterton P. 2014. *Low Impact Living: A Field Guide to Ecological, Affordable Community Building.* Earthscan, London.

1.6 (continued — right column top)

8. Habraken, N. J. 1972. *Supports: An Alternative to Mass Housing.* Translation from Dutch by B Valkenburg ARIBA. The Architectural Press, London.
9. Brand, S. 1994. *How Buildings Learn: What Happens After They're Built.* Viking, London.
10. Kendell, S. & Teicher, J. 2000. *Residential Open Building.* Spon Press, London.
11. Egan, J. 1998. *Rethinking Construction.* Department of Trade and Industry, London.
12. Ball, M. 1996. *Housing and the Construction Industry: A Troubled Relationship.* Policy Press, Bristol.

2.1 COMFORT AND SUFFICIENCY

1. Fanger, P. O. 1970. *Thermal Comfort: Analysis and Applications in Environmental Engineering.* Danish Technical Press, Copenhagen.
2. de Selincourt, K. 2024. Handled with Care. *Passive House Plus Magazine.* Issue 46, UK edition, Blackrock, Ireland.
3. de Dear, R. 2006. Thermal counterpoint in the phenomenology of architecture – A Psychophysiological explanation of Heschong's 'Thermal Delight'. The University of Sydney, Australia.
4. Heschong, L. 1979. *Thermal Delight in Architecture.* MIT Press, Cambridge, Massachusetts.
5. Alter, L. 2021. *Living the 1.5 Degree Lifestyle: Why Individual Climate Action Matters More Than Ever.* New Society Publishers, British Columbia, Canada.

2.2 CLOSING THE PERFORMANCE GAP

1. Quoteresearch. In theory there is no difference between theory and practice, while in practice there is. *Quote Investigator*. 14 April 2018. quoteinvestigator.com/2018/04/14/theory [accessed 21/10/24]
2. Siddall, M. 2022. *Thermal Bypass Risks; a Technical Review. Passivhaus Trust, London*.
3. de Dear, R. 2006. Thermal counterpoint in the phenomenology of architecture – A Psychophysiological explanation of Heschong's 'Thermal Delight'. The University of Sydney, Australia.

2.3 FORM FACTOR, MASSING AND SHAPE

1. Romm, J. 1994. *Clean and Lean Management: How to Boost Profits and Productivity by Reducing Pollution*. Kodansha, New York.

2.4 ENVIRONMENTAL MODELLING AND TARGETS

1. Lewis, S. 2017. *PHPP Illustrated: A Designer's Companion to the Passivhaus Planning Package*. RIBA Publishing, London.

2.5 PASSIVE SOLAR DESIGN

1. Holladay, M. 2010. Solar Versus Superinsulation: A 30-Year-Old Debate. *Green Building Advisor*. greenbuildingadvisor.com/article/solar-versus-superinsulation-a-30-year-old-debate.

2.6 WHY PASSIVHAUS?

1. Feist, W. Personal communication.
2. Grant, N. 2008. *A Critique of the CSH Water Efficiency Requirements*. Good Homes Alliance, London.
3. Grant, N. & Siddall, M. 2015. *Claiming the Passivhaus Standard*. Passivhaus Trust, London.
4. Charles Eames interview 'What Is Design?' (1969) in *Eames Design: The Work of the Office of Charles and Ray Eames*, eds. John Neuhart, Marilyn Neuhart and Ray Eames (New York: Harry Abrams, 1989), 14–15.
5. Cotterell, J. and Dadeby, A. 2012. *The Passivhaus Handbook: A Practical Guide to Constructing and Retrofitting Buildings for Ultra-Low Energy Performance*. Green Books, London.
6. Baeli, M. 2025. *Residential Retrofit: 24 Case Studies*. RIBA Publishing, London.

7. de Selincourt, K. & Clarke, A. The Right Time for Heat Pumps in Retrofit: Decarbonising Home Heating in a Staged Retrofit. *Passivhaus Trust*, April 2024. passivhaustrust.org.uk/UserFiles/File/research%20papers/Heat%20pumps%20and%20Retrofit%20v1.2%20240605.pdf.
8. Erskine, R. Insulate Britain! Yes, But By How Much? *EssaysConcerning*, 8 November 2021. essaysconcerning.com/2021/11/08/insulate-britain-yes-but-by-how-much.

3.1 HEAVYWEIGHT OR LIGHTWEIGHT?

1. Hewitt, M. & Telfer, K. 2012. *Earthships in Europe*. BRE Press, Garston.
2. Passipedia. 2022. Insulation vs. thermal mass. passipedia.org/planning/thermal_protection/thermal_protection_works/thermal_protection_vs._thermal_storage.
3. Architype. Ysgol Trimsaran. architype.co.uk/project/ysgol-trimsaran.

3.2 UPFRONT CARBON EMISSIONS

1. Alter, L. 2024. *The Story of Upfront Carbon: How a Life of Just Enough Offers a Way Out of the Climate Crisis*. New Society Publishers, Gabriola Island, Canada.
2. Pearson, C. & Waters, E. 2023. The Missing Embodied Carbon Link: Construction. *BuildingGreen*. buildinggreen.com/feature/missing-embodied-carbon-link-construction.
3. Clarke, A. & Grant, N. 2010. Biomass – A Burning Issue. *AECB*. aecb.net/download/biomass-a-burning-issue.
4. Alter, L. 2021. How the Serpentine Pavilion Can Save The Planet. *Treehugger*. treehugger.com/serpentine-pavilion-claims-carbon-negative-not-true-5189683.
5. ZSL. Living Planet Report. zsl.org/what-we-do/projects/living-planet-report.
6. Alter, L. 2023. How Good is Wood? *Green Building Advisor*. greenbuildingadvisor.com/article/how-good-is-wood.
7. Woodscape. Natural Durability Classification of Timber. woodscape.co.uk/natural-durability-classification-of-timber
8. Koenig G. 2015, Eames, Taschen, Cologne.

3.5 SERVICES

1. Hawkins, G. 2011. *Rules of Thumb: Guidelines for Building Services*, 5th edn.. BSRIA, Berkshire.
2. Hazard, P. Essential Lessons From Electric Schools. *CIBSE Journal*, March 2024. London.

4.1 LAND

1. Shrubsole, G. 2019. *Who Owns England?: How We Lost Our Land and How to Take It Back*. Harper Collins, London.
2. Moore, R. 2023. *Property: The Myth That Built the World*. Faber & Faber, London.

4.2 DESIGN EDUCATION AND POST-OCCUPANCY EVALUATION

1. Al Shawa, B. The twisted Education of Architects. misfitsarchitecture.com/2011/06/04/the-twisted-education-of-architects.

4.3 COST AND VALUE

1. Axon, C. J. & Roberts, S. H. Analysing the Rising Price of New Private Housing in the UK: A National Accounting Approach. *Habitat International* 130, December 2022. Elsevier B.V.

4.4 VALUE ENGINEERING

1. McLeod, R. 2024. Will better environmental standards really mean fewer homes? *The Herald Scotland*, 13 May 2024.
2. Rt Hon Sir Oliver Letwin MP. 2018. *Independent Review of Build Out*. Ministry of Housing, Communities and Local Government, London.
3. UK Parliament Research Briefing. 2023. *Tackling the Under-supply of Housing in England*. UK Parliament, London.

4.5 APPOINTING A DESIGN TEAM AND CONTRACTORS

1. Segal W. 1977. Address to the Royal Institute of British Architects.
2. Brinkley, M. 2023. *Housebuilder's Bible*, 15th edn. Ovolo Books, Cambridge.
3. Curtis, R. Grenfell Inquiry has Exposed Design and Build as Our Dirty Little Secret. *Architects' Journal*, 22 February 2020. EMAP, London.

Acknowledgements

Good building is a process that involves many people from many fields. This can make it difficult to give credit where credit is due as many of the best ideas come out of multiway discussions and a long process of trial and error.

If you are reading this, you are probably checking to see if you got a mention and so you probably deserved one even if our aging brains managed to forget you. The following is a non-exhaustive list of a few of the many friends and colleagues who have influenced our work over the years. It includes a mix of builders, designers, engineers, scientists, clients, even a few journalists. Some we have known for all our working life whilst others managed to change our direction of travel in a single chance meeting.

In a very particular order they are:
Ken Atkins, Bill Bordass, Bronwyn Barry, Mark Barry, Sam Brown, Bill Butcher, John Cantor, Jeff Colley, Colin Chetwood, Wolfgang Feist, Sally Godber, Charles Grylls, Sandy Halliday, Cindy Harris, Chris Herring, Michael Herring, Jono Hines, Bjørn Kierulf, Howard Liddell, John Little, Bob Lowe, Neil May, Juraj Mikurcik, Mark Moodie, David Olivier, Tim Padfield, Dai Rees, Brian Richardson, Kara Rosemeier, Walter Segal, Tahir Sharif, Mark Siddall, Ted Stevens, John Turner, Colin Ward, Peter Warm, Mike Whitfield, Beth Williams.

We are particularly grateful for those who took the time and care to provide technical and editorial feedback on the text: Lloyd Alter, Pat Borer, Fran Bradshaw, Alan Clarke, Sarah Lewis, Carl Meddings, Kate de Selincourt and our very patient editor at Bloomsbury, Heather Bradbury. Also, Nicola Liddiard for the inside design work, Amanda Keyte for the cover, Beth Dymond for the copyedit and Jacky Ferneyhough for the proofread and index.

The following friends very much influenced our writing but sadly did not live to see the final result. Howard Meadowcroft's beautiful hand drawn details always integrated function and beauty. John Willoughby brought his razor-sharp wit and scepticism to many discussions and Monte Paulson was writing a book developing a modern-day pattern language for sustainable buildings that we hoped would be the follow-on practical manual that we have only hinted at. They all really cared about what they did.

Nick's partner Sheila is a potter from a family of builders and she designed and built at least half of their house from the drains upwards. At the age of 69, her brother has just started building his first Passivhaus with his son's family. Jon's partner Rona Nicholson was an exacting client for their house, has in-depth experience of UK housing policy and practice and has been keeping the show on the road throughout.

Image credits

Bloomsbury Publishing would like to thank the following for providing photographs and for permission to reproduce copyright material. While every effort has been made to trace and acknowledge all copyright holders, we would like to apologise for any errors or omissions and invite readers to inform us so that corrections can be made in any future editions of the book.

All inside photos and diagrams except the below are copyright Jon Broome and Nick Grant.

Key: Left = L, Right = R, Top = T, Middle = M, Bottom = B, SS = Shutterstock, F = Flickr, A = Alamy, W = Wikimedia

Insides: Lucy Casson: 3, 41T. Tijmen Kattevilder: 10–11. Juraj Mikurčík: 12, 31, 58, 108, 143, 145, 150, 168, 186, 198, 225T, 225B. RIBA/Louis Hellman: 14, 169. SS/Walencienne: 16. F/Dullhunk: 17. A/Dennis Gilbert: 20. A/Imaginechina Limted: 24. Tim Padfield: 29T. Morten Ryhl-Svendsen, Poul Klenz larsen, Lars Aasbjerg Jensen and Tim Padfield: 29B. Imperial War Museums: 32. National Records Scotland: 35T, 35BL. A/Place:Image: 35BR. S/Oliver Foer: 36. NG, with permission National Records Scotland: 38T, 38B. Wikimedia Commons: 41B. A/OBNW: 42. Adelina Iliev: 43. SS/Milos Ruzicka: 44. A/Steve Parsons: 49. A/Richard Hilsden: 51. SS/Adwaith: 52. DUG Wilders/HEM Architects: 54. SS/Gherzak: 64. A/Stefano Ravera: 65L. Ray Main © Mole Architects: 70. James Durham: 72. Tom Bird: 80. Cullinan Studios: 81. A/geogphotos: 85. Levitt Bernstein Architects: 97. Sam Simmons: 98. Lev Kerimol: 101. RUSS: 102B. Solk Photography: 106, 160. Ros Kavanagh: 110. Jennifer Crawford: 111. Alex Gregory (www.CartoonStock.com): 113. Jim Stephenson © Mole Architects: 118. A/Christopher McGowan: 121. Sally Godber, Coaction: 122L, 122R. W/Marco2000: 123. James Thornbery Creative Commons Attribution-Share Alike 4.0 International: 124. David Butler Photographer © Mole Architects: 127. Taran Wilkhu: 129, 277T. SS/Claudio Divizia: 130. A/Simon Kirwan: 132. Lisa Logwig: 134. Graeme Deas: 144B. Jack Hobhouse: 146. W/Christopher Tweed: 148B. LEAP_MarkSiddall: 149. David Barbour Photography: 151. Dawn Keyse: 152. Kimbo Fidelo Sito, Levitt Bernstein Architects: 155. Passivhaus Trust: 158. Iain Richardson: 161TL. Glyn Hudson: 161R. Tom Hargreaves: 163. Simmonds Mills and Hannah and Otis Sloan Wood: 164, 166T. Photo DB Plaster: 167T. DB Plaster 167B. EcoCocon: 170–171, 182. Will South, Etude: 172. Phil Boorman Photography Ltd: 174. Chryssa Thoua: 175. David Barbour Photography: 190. Porotherm UK: 193T. Mark Siddall, LEAP: 215. A/imageBroker.com: 218. SS/Seven Hansche: 228. Google Maps: 232. Photo Jack Hobhouse: 244. SS/Jessica Girven: 258. G/Grace Robertson: 260. A/World History Archive: 262. SS/Janet Ebner: 265. Amsterdam City Archive: 268. W/Balon Greyjoy: 271. W/Hesham Ferouk Ragab: 272–273. Sheila Herring: 277B.

Front cover and spine: A/Daryl Mulvihill.

Bold pagination refers to the main entry for a topic; *italic* pagination refers to images.

Index

A

access
    construction 204, 208
    garden 121
    maintenance 186, 210, 211, 224, 225
    site 200, 203, 232
accessibility 17, 157
accessible construction 76, 86, 196
Adamson, Bo 147
adaptability **80**, 81
    design/construction 25, 53, 69, 76, 88, 102, 187, 203, 237
    longevity 69, 73, 97, **186**
    see also prefabrication
Aeschylus 133
aesthetics ('beauty') 8, **18**, 23, 41, 54; *19*
    defining 15, 16, 24
    design detail 143, **194-95**, 212
    planning system 24, 89, 213, 265, 269
    see also pattern language; planning system
airtight design/construction 116, 133, **165-67**, 191, 192, 193, 195, 203, 204, 208, 209, 248-49; *116, 192, 193*
    measuring 166-67, 251
    roofs 210, 212; *210*
    Segal method 89, 196, 200
    windows 216, 248, 250; *217*
    see also thermal bridge; ventilation
Alexander, Christopher 15, 66, 67, 68, 142
    see also pattern language
Alter, Lloyd 45, 111, 245
American Air Museum (Duxford) 27
Anderson, Jane 188
Architectural Registration Board 238
Architype 31, 33; *10, 88, 89, 98, 102, 106, 146, 60, 226, 244, 257*
archive and museum building **23-25**, 35, 222; *24, 35*
    case studies 26-29, 31, 33-34, 35; *26, 28, 29, 30, 32, 34*
    Passivhaus 163, 220

Arnstein, Sherry 74
Atkins, Ken 76
Axon, Colin 242

B

Bacon, Richard 90
Ball, Michael 79
basements 177, 199
Beck, Harry 130
Bilbao Effect 15
biodiversity 96, 188, 212, 229, 230, 270
biomass heating 51, 157, 199
Biomimicry 48
Brand, Stewart 78
Brewster, Benjamin 115
Brinkley, Mark 256
building management systems (BMS) **220-21**, 225, 243
    see also services
building on grade 201
building regulations UK 111, 126, 169, 176, 181, 204, 266
Building Research Establishment (BRE) 61
Building Safety Act 267, 270

C

carbon
    capture and storage 184
    measures 111, 180
    net zero/neutral claims 46, 47, 157, 158, 159, **183-85**
    offsetting 179, 183, 184
    targets 157, 179, 243
    upfront 120, 121, 125, 126, 157, 159, 161, **179-87**, 188-89
    see also sustainability (whole life); upfront emissions
carbon, low
    construction 120, 154, 216, 250
    heating 51, 136, 138, 139, 158, 162, 181
    materials 69, 122, 185, **187-89**, 203
    see also Passivhaus standard
Carbon Trust 115
Casson, Mick 113
cathedral ceiling 210, 211; *211*
Centre for Alternative Technology (Wales) *234*
Charles III 65, 86
Chartists 85

Chinese State Council 41
Church of England 85, 229, 232; *231*
circulation spaces 38, 57
circulation, air *see* ventilation
cladding 34, 89, 189, **205**, 215, 216, 248; *34, 72, 192, 201, 204, 217, 226*
    see also Grenfell Tower
Clarke, Alan 162, 181, 220
Code for Sustainable Homes (CSH) 126, 127, 147-48
Collinson, Andrew 203
comfort, thermal **107, 109**
community land trust (CLT) **98, 99**, 102, 233, 271; *100*
    RUSS Community Land Trust 102-03; *10, 76, 102*
community-led housing and building projects 42, 71, 83, 84, 97, **98-101**, 154, 232, 269
    Community Infrastructure Levy (CIL) 90, 231
    Community Led Homes 99, 255
    co-operative development 66, 97, **98, 99**, 206, 255; *94, 101*
    finance and funding 86, 90, 99, 100, **101**, 103, 162, 233
    risk and responsibility 255, **264-65**
    see also self-build
competitions, architectural 23, 45, 119, **255**
composting toilets 51
concrete construction 154, **173**, 177, 184, 188, 193, 212; *204*
    avoiding/reducing 200, 203; *203*
    frames 78, 81, 103, 122, 204, **209**; *122, 204*
    heat storage 133
    pad foundations 86, 200, 201, 202; *200, 201*
    Passivhaus 154, 157
condensation 96, 116, 166, 211, 216
    see also mould
Construction (Design and Management) Regulations (CDM) 267
cooling 26, 131, 158, 221
    night 143, **175, 176**, 214, 225; *142*

cork 46-47, 184, 189
Cork House (Berkshire) 46-47
cost and economics **241-43**
    affordability 15, 16-17,
        18-19, **21**, 42, 64, 71, 83,
        101, 102
    see also finance and
        funding; value
        engineering
costs, building 46-47, 68, 119,
        167, 191, 204, 219, 223,
        237, **241-43**, 250, 251
    design budget 34, 46-47,
        53, 55, 119, 154, 200, **241**
    Segal/self-build approach
        69, 83, 88, 89, 101, 103,
        195, 196, 200
Crawford, Jennifer 111
cross laminated timber (CLT)
        188, 193, 212; *202, 204,
        208*
Crown Estate 85, 229; *231*
Curitiba (Brazil) 94
Curtis, Russell 258

D

Dayes, Kareem 102
Dear, Richard de 109
Delft self-build (Netherlands)
        84; *74*
Den Bosch spherical house
        (Netherlands) *44*
Den Haag self-build/
        community (Netherlands)
        84, 233; *82*
Denmark co-housing 98
design
    brief 242, 243
    user input 25, 42, 53, **66**,
        75, 81, 93, **97**, 273
design and build (D&B)
        procurement 90, 241,
        247, 251, **257-59**
design detail 19, 122, **191-96**,
        248, 250, 251, 261
    aesthetic 143, **194-95**
    external envelope **191-92**,
        204, 212, 249; *192*
    general principles 186, 188,
        195-96
    windows and doors 189,
        215, **216-17**
    see also pattern language
design education 235-39
design process **53-56**, 246
    adding detail **55**, 99, 257,
        258
    circular process 54-55

design team **54**, 56
    fundamentals **56-57**, 144
developers 83, 97, 270
    commercial (large) 42, 99,
        101, 230, 231-32, 233
    small 42, 90
Devonport Dental Hospital *41*
Diggers, The 85
Dovehouse Court (Solihull)
        127
drainage
    fenestration 216
    plumbing layout 224, 225
    roofs 210, 211, 212
    surface water 31
drawings and sketches, value
        of **60**, 87, 195-96, 236;
        61, 87
Duffy, Frank 78
durability 69, 97, 185, **186-87**,
        212, 237
    see also maintenance;
        sustainability (whole life)
dweller control **66**, 71, **75**
    see also design, user input
Dyckhoff, Tom 39
dynamic modelling **131**, 173,
        175

E

Eames, Charles 18, 23, 45,
        150, 189
Eames, Ray 18, 23, 150
Earthships 173; *172*
EcoCocon panels 181; *150, 170,
        182*
EcoHomes 147, 148
Edward Cullinan Architects 81
Edwards Court (Exeter) 107
Egan, John 79
Elytra Filament Pavilion
        (London) *48*
enabler role 97, 99, **255**, 265
energy performance
        standards 83, 91, 96, 101,
        117; *240*
    see also EnerPHit standard;
        Passivhaus standard
energy, low (design/building)
        26, 31, 83, 90, 128, **177**,
        202, 222, 233
    glazing 214, 215-16; *216*
    Passivhaus 96, 131, 149,
        154, 161, 250
    see also Passivhaus
        standard
EnerPHit standard 160, 161;
        80, 149

    see also Passivhaus
        standard
environmental modelling
        125-31
Erskine, Richard 162
Europe 229, 270
    North European housing
        policy 97, 98, 255, 265
    self-build 84, 128, 233
    timber 188
Eyck, Aldo van 268

F

Fanger, Povl Ole 79
Farnsworth House (Illinois) 18,
        123; *123*
Feist, Wolfgang 115, 147, 149,
        154, 163
fenestration, design impacts
        of 133, 144, **248**, 250
    see also windows
fill/infill, wall 130, 201,
        203-04; *117, 202, 203*
    see also insulation
finance and funding 24, 61,
        66, 231, 237, 243, 265
    self-build/community 86,
        90, 99, 100, **101**, 103, 162,
        233
    see also cost and
        economics; costs,
        building
Fire Safety Act and
        regulations 237, 265,
        267
fire safety and construction
        57, 117, 121, 125, 130, 157,
        204
    archive buildings 25, 26,
        29, 31, 33, 34
    material performance 177,
        189, 208, 209
    see also Grenfell Tower
foam glass gravel 203
form factor **120-21**, 128,
        129-30, 142, 159, 194,
        210, 247; *120*
fossil fuels 73, 111, 138, 158,
        181, 184
Foster, Norman 27
foundations 185, 199, **200**,
        201, 248; *201, 203*
    Segal design/build 57, 86,
        88
frames 206; *206*
    concrete 78, 81, 103, 122,
        204, **209**; *122, 204*

post-and-beam 69, 76, **78**, 200; *200*
steel 21, 34, 78, 122, 154, 157, 177, 193-94, 204, **209**; *80, 193, 204*
timber 86, 89, 188, 192, 193, 200, 201, 204, 206, **209**; *117, 151, 198, 248*
framing, balloon 206, 207, 208, 248
functionality 17, **21**, 123, 224, 242, 246; *19*
and form 18, 25, 54, 194, 195
see also pattern language; utility

G

Germany
community housing 95, 98, 99, 255, 265; *94*
self-build 84, 128
Girardet, Herbert 93
glazing 129, 143, 181
design/managing large expanse 21, 133, 134, 135-37, **138, 140**, 154, 176, 214, 248; *20, 22*
doors 143
low-energy 214, **215-16**; *216*
triple 109, 136, 137, 140, 141, 214, 215, 248
window design 213-14
see also radiant asymmetry
glulam 188, 242
Godber, Sally 122
Grant, Mhairi *151*
green roof 88, 94, 103, 212; *88, 212, 261*
Grenfell Tower 61, 103, 117, 263, 265, 267, 270
inquiry 117, 258, 263, 265
Guangzhou Circle (China) 41; *41*

H

Habraken, Nicholas John 78
Hackbridge School (England) *146*
Hackney, Rod 86
Hadid, Zaha 37
Hawkins, Glenn 219
Headway Gardens (Walthamstow) *91*
Health and Safety at Work Act 267
heat pumps **161-62**, 220, 224-25
heating

biomass and wood pellet 51, 157
fossil fuels 138, 158, 162, 223
gas, synthetic 158
non-domestic buildings 220, 223
photo voltaic (PV) 138, 158, 159, 225
see also Passivhaus standard
heat recovery (MVHR) 110, 117, 134, 139, 140, 148, 169, 181
heavyweight construction **173-77**
see also concrete
Hereford Archive and Records Centre 31
Heschong, Lisa 109
Holladay, Martin 134
Hope View House (England) 135, 138, 176; *134, 176*
hospitals 61, 121, 219; *15, 41, 81*
Passivhaus standard 157, 163
hot water system rectangle 223-24; *224*
Hubner, Peter 81
humidity 29, 141, 157, 165, 169, 222; *29, 35*
see also archives and museums; condensation
Hunstanton School (Norfolk) 21; *20*
hygrothermal modelling 29

I

I beams 128, 201, 209; *201, 202, 203, 204, 208, 211, 248*
Illinois Institute of Technology 37
Imperial War Museum (Duxford) 33
Indian Coffee House (Trivandrum) *52*
Industrial Revolution 65, 74
inefficiency, place for **113**
infrastructure
professional 90, 97, 265
public 181, 231, 233, 269, 270
insulation 139, 153, 175, 177, 181, 205, 248
cork 46, 47
EPS/XPS 202, 203, 211; *205*
external wall (EWI) 34, **209**; *201, 202*

floor 201-03
foam 116
roof 194, 210, 211; *193, 211*
superinsulation 133, 134
wall **129, 130**, 160-61, 162, 166, **192-94**, 200; *192, 193, 200, 202*
see also thermal bridge; thermal bypass
integrated design 196, 221-22, 251, 258

J

Jensen, Poul 26
Jessop Hospital (Sheffield) *15*
Jevons, William Stanley 111
joists 188, 210; *193, 208, 210*

K

Kettle, Tony 28
Klein, Gary 223; *224*
Krier, Leon 65

L

Lambeth Community Care Centre 81
land ownership 85, **229-33**
barriers 90, 99
finding a site **232-33**, 270
values 111, 230, 231, 242, 245-46
Lewisham Self-Builders 76, 86, 102, 267, 268; *77, 86*
Libeskind, Daniel 39; *40*
Liddell, Howard 45
lightweight construction **173-77**
see also Segal design/build; timber construction
listed buildings, performance of 14-15, 18, 21
Lloyd Wright, Frank 18; *124*
Lloyd's Building (London) 37
Local Agenda 21 80
local authority
land 233
regulatory authority 265, 266
Local Government Association 186
London Metropolitan University 39; *40*
London Underground 230; *130*
Lyons, Eric *247*

M

Mahatma Gandhi 113
maintenance 89, 188, 189, 191
    accessible design 28, 38,
        186, 211, 224, 225, 237
    costs 24, 99, 101, 215, 243
    documentation 268
    services 51, 221
McKay, Graham 8, 235, 275
mechanical and electrical
        (M&E) services 25, 54,
        68, 219
Melfield Gardens (Lewisham)
        97
Menges, Achim *48*
Mikurčík, Juraj 154, 195; *58,
        108, 150*
Miles, Lawrence 246
minimalism
    architectural/design 123,
        195
    eco-minimalism 45
    engineering 123
    Passivhaus design 154
Miškov, Roman 59
mixed use development 95,
    97
Mole Architects *70, 118, 127*
monolithic construction 194,
        209; *193*
Moore, Rowan 231
mould 51, 96, 109, 117, 154, 161,
        162, 165, 166
    see also condensation
museum *see* archive and
        museum building
Mutual Home Ownership 97

N

Nanning Library (South China)
    24
National House Building
        Council 189
Natural Cork Council 47
Netherlands *44, 95, 214*
    Delft self-build 84; *74*
    Den Haag self-build/
        community 84, 233; *82*
    housing approach 98, 233,
        255
    Oosterwold self-build 84;
        84
New Brutalism 21, 122
North America building trends
        84, 98, 133, 177, 185, 203

O

Olivier, David 202
Oosterwold (Netherlands) 84;
    84
open book contracting 259
open spaces 95-96
Orr, Harold 128

P

Padfield, Tim 26, 29, 31
Paper Igloo *151*
Parnell, Stephen 184
Passivhaus Institute 115, 117,
        153, 154, 158
Passivhaus Planning Package
        (PHPP) **131**, 135, 137, 140,
        153, 154, 167, 176, 250
Passivhaus standard 125, 129,
        163, 166, 175, 273
    airtightness and insulation
        134, 166, **248**
    benchmark for efficiency
        111, 115, 120, 122, 147, 148,
        149
    constraints and challenges
        150, **153-54**, 157, 163
    form factor **120**, **122**, 125,
        129, 130, 247-48
    hospitals 157, 163
    materials 181, 193, 194, 202,
        203, 209; *182, 193*
    peak heating load 135, 137,
        138-39, 167, 181; *138*
    Plus and Premium **158-59**
    practical examples 136,
        137-38, 154, 220; *12, 43,
        58, 106, 108, 127, 137, 148,
        150, 151, 152, 155, 156,
        163, 191, 226, 244*
    retrofit **160-61**, 162; *160, 161*
    schools 149, 157, 163, 176,
        202, 220, 223; *146, 174*
    value engineering **250**
    see also concrete
        construction; glazing,
        low-energy; steel
        construction
Passivhaus, Scotland 148,
        209
pattern language 15, **66-68**,
        76, 88, 129, 142, 250
Paulsen, Monte 67-68
photo voltaic (PV) heating
        138, 158, 159, 225
Pieksma, Henk 29
piles and piling 200, 203; *200*
Pirsig, Robert 21

planning system
    appearance (aesthetics)
        24, 89, 213, 265, **269**
    control **231-32**, 247, **269-71**
    Europe 233, 270
    issues 79, 90, 93, 103, 111,
        230, 253
plastic and alternatives 185,
    189
plastic waste 187, 189
plastic, glass-reinforced 215
plyscrapers 188
Pompidou Centre (Paris) *218*
post-occupancy evaluation
        157, **238-39**
post-war construction
    First World War 85, 231
    Second World War 65, 85,
        231, 246, 269
Poundbury (Dorset) 65
Pre-Construction Services
        Agreement (PCSA) 259
prefabrication 89, 185; *185*
    manufacture 78-79, 81,
        206; *150, 208, 251*
primary energy renewable
        (PER) 158
Private Finance Initiative 243
procurement
    contractors 256-59
    cost 119
Public Services (Social Value)
        Act 99, 233
PVC 214-15; *200*
    alternatives 189, 213, 215

Q

qualities, good building **14**
    assessing 14-17
quantity surveyor (QS) 242,
        254

R

radiant asymmetry 107, 109,
    140-41
raft slab 201-02
rainwater harvesting 51
Read, Carveth 131
reclaimed materials 285; *156,
        186*
refurbishment and reuse 161,
        182, **185**, 186-87, 237; *80*
    see also upfront emissions
resources, sufficiency and
        consumption of 69, 83,
        93, 96, 111, 183, 187
retrofitting 18, 54, **160-62**,
        166, 182

Right to Build
   policy 42
   register 233
   task force 83
Right to Buy policy 42
Rio Earth Summit 80
Roberts Limbrick 34
Roberts, Simon 242
Rogers, Richard 37
Rohe, Ludwig Mies van der
   18, 37, 123; *123*
Romm, Joseph 119
roof construction **210-12**; *210,*
   212
   CLT 188, 189
   insulation 130, 161, 167,
      193-94, 210, 211; *116, 192,*
      193, 211
   rafters 19, 249 *211, 249*
   ventilation (breathability)
      166, 210, 211; *116, 210, 212*
   wall junctions 122; *122*
roof design
   details 186, 210-12; *211*
   examples 28, 29, 33, 128,
      159, 195, 215; *159*
   green 88, 94, 103, 212; *88,*
      212, 261
   Segal method 56, 89, 196
rooflights 142
Royal Institute of British
   Architects (RIBA) 42,
   86, 238
rules of thumb calculations
   129, 130-31, 166, 192, **219**,
   250
Rushby, Kirk 203; *12, 199*
RUSS Community Land Trust
   102-03; *10, 76, 102*

S

Saint, Andrew 13
Sanford Housing Co-op
   (Lewisham) *101*
Saskatoon Conservation
   House 128
Scandinavian housing
   development 97, 98, 255
schools
   Austria/Germany 153
   building standards 61, 117
   design 81, 121, 140, 214
   examples 21, 176 *20, 132,*
      146, 174
   modelling 125, 135, 219
   Passivhaus 149, 157, 163,
      176, 202, 220, 223; *146,*
      174

Schroeder House
   (Netherlands) *214*
sculptural architecture 26,
   28, 37-41, 235; *24, 26,28,*
   36, 38, 40, 41, 42, 44,
   218
sculpture *41, 271*
Segal design/build
   energy performance 57,
      69, **89**
   legacy 102, 185
   method **69**, 87-88, 89,
      242; **56, 77, 86, 87, 88**
   non-builders **76-77**, 86,
      195-96
   post-and-beam frame 69,
      76, 78, 200
Segal, Walter 57, 76-77, 86,
   89, 242, 254, 269
self-build 42, 66, **83-91**, 257,
   267; *63, 150, 161, 177*
   Australia 84
   costs, building 69, 83, 88,
      89, 103, 195, 196, 200,
      260
   funding 86, 90, 99, 100, **101**,
      103, 162, 233
   opportunities and barriers
      (UK) 85, **90**, 97, **232-33**
   see also Segal design/build;
      Lewisham Self-Builders
Selincourt, Kate de 162,168
Serpentine Pavilion (London)
   184
services **219-25**
   access 186, 225
   design stage 54, 55, 78,
      220; *206*
   exterior feature 37; *218*
   installation 191, 192, 194,
      209; *193, 210*
   integrated design 54, 220,
      221-22
   M&E 25, 54, 68, 219
   management systems
      (BMS) **220-21**, 225, 243
   principles 225
Shrubsole, Guy 229, 230
Siddall, Mark 115; *149*
Smithson, Alison and Peter 21
social housing 76, 100, 126,
   233; *14, 240, 242*
social values 99, 233, 275
solar design, passive **133-44**,
   176, 224
   superinsulation 134, 181
   see also radiant asymmetry
solar gain 135-36, 139, 142

Somerset Heritage Centre
   (Taunton) 34; *34*
space sufficiency **110-11**
Spreefeld (Berlin) 66, 206;
   206, 207
Springhill Co-housing (Stroud)
   98; *98*
St George's Hill (Surrey) 85
St George's School
   (Merseyside) *132*
stack ventilation 51, 225
steel construction 21, 78, 177,
   184, 185, 188, 202, 209;
   80
   insulation/thermal bypass
      34, **193-94**, 204, 205,
      208, 212; *193, 204, 205,*
   Passivhaus 154, 157, 209
   piling 200; *200*
   reusable 186
straw building/insulation 181,
   184-85, 211
   bales 128, 129,177, 194; *58,*
      178, 180
   EcoCocon panels 181; *150, 170,*
      182
   stud frame/wall (timber/I) 88,
      89, 116, 117, 167, 192, 193,
      209, 248; *202, 204, 208*
Suffolk Record Office
   (Ipswich) 29, 31
sustainability (whole-life) 46,
   53, 81, 157, 179, **180**,
   181-82, 185, 186-87, 243;
   215
   see also carbon; durability;
      upfront emissions
sustainable neighbourhoods
   93-103, 233
   open spaces **95-96**
   mixed use development 95,
      97
   traffic **95**
   see also community land
      trust; community-led
      housing and building
      projects
Switzerland housing
   development 255

T

target cost contracting 259
technology and AI 48, 122
tendering contracts 53, 242,
   243, 254, 259
   competitive tendering 54,
      56, 255, 257
terraced housing 121; *118, 120*

thermal bridge **117**, 131, 134,
154, 175, 250; 135, 138
  avoiding 194, 201, 202,
205, 209, 212
  causes 193-94, 205, 208,
212, 248
thermal bypass 115, **116**, 130,
131, 175, 199
avoiding 192, 209
causes 117, 205, 210
thermal mass 26, 173, 175,
176, 177
timber
  forestry industry 188, 215
  low carbon/sustainability
126, 157, 179, 183, 184-85,
187-88; *215*
  treatment 186, **188-89**
timber construction 177, 211,
212
  frames 86, 89, 192, 193,
200, 201, 204, 206, 209,
248; *117, 151, 198, 248*
  heat loss 117, 176
  see also cladding; Segal
design/build
toilets and waste treatment
51, 93
Town and Country Planning
Act 231, 269
Townsend, Geoffrey *247*
Trading Standards 149
tubes, earth 51
Turner, John 66, 75

U

upfront emissions 34, 111, 125,
126, 136, 161, **180-87**,
188-89
  form factor **120-21**, 127, 157,
159
  measuring 173, **179-80**
Upton, Polly 203; *12, 199*
Usable Building Trust 115
utility 16, **17**, 18-19, 24, 25, 34,
131, 273
  see also functionality
U-value 116, 117, 128, 131, 175,
204, 250; *120*

V

value engineering (VE) 119,
205, **245-47**, **250-51**
  barriers 247
vapour control 116, 166, 210,
211; *116, 208*
  see also airtight design;
condensation; humidity

Vauban (Freiburg, Germany)
95, 265; *94, 104, 264*
ventilation **165-69**, 220, **224**;
142
  heat recovery (MVHR) 110,
117, 134, 139, 140, 148,
169, 181
  mechanical 117, 157, 162,
168, **169**, 221, 225
  natural 143, 157, **168**,
213-14, 225
  passive stack 51, 143, 165
vernacular
  modern 48, **63-69**, **71**
  practical construction
73-81
  sustainable 69, 71, 73, **74**,
75, **80-81**, 250, 273
  traditional 19, 64, 69
Vitruvius 15

W

walls
  AAC block 201, 202, 203,
209; *201, 203*
  design and construction
177, 188, 196, **203-04**,
206, 208-09, 250; *211,
248*
  heat loss and insulation 115,
116, 122, 160-61, 166, 167,
192-94, 200; *192, 193,
200, 202*
  insulation thickness 129,
130
  living walls 224
  Segal method 76, 89
  services 191, 192, 194, 209,
225
  see also cladding; framing;
insulation
Ward, Colin 66, 75
Warren Benbow Architects *see*
Hope View House
waste recycling/reduction 93,
94, 189, 243
water
consumption 51, 80, 93, 94,
103, 147
design feature 29, 41
energy 94, 103, 133
hot water systems 133, 153,
222-24, 225; *224, 225*
supply 51
toilets 51
  see also condensation;
humidity

water ingress 201, 211, 212,
217; *216*
  flat roofs 28, 211
Whitfield, Mike 249
wildlife *see* biodiversity
Wilton, Oliver *46*
wind
  designing for 34, 192-93,
209, 214, 249; *192, 193,
208*
  stack ventilation 51, 225
  turbines 51, 138, 147
  see also airtightness;
thermal bridge; thermal
bypass
wind washing 116, 210
  see also thermal bypass
windows *see* glazing
window and door frames
214-15
Woodhorn Museum and
Archives (Ashington) 28;
28

Y

Ysgol Trimsaran (Wales) 176;
174

Z

Zoological Society 188